Health Care Issues, Costs and Access Series

A Guide to Public Health Research Needs

HEALTH CARE ISSUES, COSTS AND ACCESS

The Health Care Financial Crisis:
Strategies for Overcoming an
"Unholy Trinity"
*Cal Clark and Rene McEldowney
(Editors)*
2001. ISBN 1-56072-924-4

Health Care Crisis in America
James B. Prince (Editor)
2006. ISBN 1-59454-698-4

A New Epidemic: Harm in Health
Care-How to make Rational
Decisions about Medical and
Surgical Treatment
Aage R. Moller
2007. ISBN 1-60021-884-9

Decision Making in Medicine and
Health Care
Partricia C. Tolana (Editor)
2008. ISBN 1-60021-870-9

Social Sciences in Health Care and
Medicine
*Janet B. Garner and Thelma C.
.. Christiansen (Editor)*
2008. ISBN 978-1-60456-286-6

Health Care Policies
*Linda A. Bartlette and Ida F. Lawson
(Editor)*
2008. ISBN 978-1-60456-352-8

Handbook of Stress and Burnout in
Health Care
Jonathon R.B. Halbesleben (Editor)
2008. ISBN 978-1-60456-500-3

Health Care Costs: Outlook and
Options
Raymond W. Inhurst (Editor)
2009. ISBN ISBN 978-1-60692-151-7

Medicare Payment Policies to
Physicians
Katherine V. Bergen (Editor)
2009. ISBN 978-1-60692-131-4

Comparative Effectiveness of
Medical Treatments
Peter Villa and Sophia Brun
2009 ISBN: 978-1-60741-109-3

A Guide to Public Health Research
Needs
Raymond I. Turner(Editor)
2009. ISBN: 978-1-60692-852-3

Health Care Issues, Costs and Access Series

A GUIDE TO PUBLIC HEALTH RESEARCH NEEDS

RAYMOND I. TURNER
EDITOR

Nova Science Publishers, Inc.
New York

LIBRARY OF CONGRESS CATALOGING-IN-PUBLICATION DATA
Available upon request
ISBN: 978-1-60692-852-3

Published by Nova Science Publishers, Inc. ✦ *New York*

CONTENTS

PREFACE

This book from the the Centers of Disease Control (CDC) focuses on four inter-related public health areas: healthy people across all stages of life, healthy places and communities, preparedness against infectious, occupational, environmental and terrorist threats and improved global health. Protecting and improving the health and well-being of people, communities and the nation is at the heart of public health. There are many things that contribute to our accomplishing our efforts, with science and public health research heading the list. The best possible science to protect the health and safety of Americans and people around the work is the hallmark of public health. Thus, the challenges of our increasingly complex and interdependent world require new approaches to generating and disseminating the knowledge and innovation needed to promote well-being and improve health. As a step toward fostering strategic investments in public health research and science, the CDC, in consultation with the public and a wide range of partners, has developed this comprehensive guide, Advancing the Nation's Health: A Guide to Public Health Research Needs.

The book is published by U.S. Department of Health and Human Services Centers for Disease Control and Prevention, December 2006.

FOREWARD

Protecting and improving the health and well being of people, communities and the nation is at the heart of public health. There are many things that contribute to our accomplishing our efforts, with science and public health research heading the list. The best possible science to protect the health and safety of Americans and people around the world is a hallmark of public health.

Today, the challenges of our increasingly complex and interdependent world require new approaches to generating and disseminating the knowledge and innovations needed to promote well-being and improve health. As a step toward fostering strategic investments in public health research and science, the Centers for Disease Control and Prevention, in consultation with the public and a wide range of partners, has developed this comprehensive guide, Advancing the Nation's Health: A Guide to Public Health Research Needs, 2006-2015.

In light of the challenges facing us in the coming decade, the research areas and needs described in this Research Guide extend beyond those traditionally associated with public health. To achieve practical, cost-effective policies, programs, and practices that improve health, the field of public health will need to place a high priority on interdisciplinary, cross-cutting research that facilitates innovations and helps inform many program areas. At CDC, we are aligning our work and research efforts to achieve specific Health Protection Goals that focus on four inter-related areas: healthy people across all stages of life; healthy places and communities; preparedness against infectious, occupational, environmental and terrorist threats; and improved global health. We have a great deal of knowledge to draw upon to help us achieve our goals in these areas, but as this Research Guide illustrates, significant research and science gaps remain in all areas.

At CDC, we plan to draw upon this comprehensive public health research compendium to guide our efforts. We expect the Guide to Public Health Research Needs will help us and the broader world of public health in many ways, from helping identify knowledge gaps to describing the range of critical research needs for public health programs to fostering the collaborations needed to achieve even greater success. We hope you find this to be an invaluable tool and that you'll help us in our efforts to keep this a dynamic resource that reflects the priority research needs of public health.

Julie Louise Gerberding,
MD, MPH Director, Centers for Disease Control and Prevention
Administrator, Agency for Toxic Substances and Disease Registry

Chapter I

OVERVIEW

A. BACKGROUND

The U.S. Centers for Disease Control and Prevention (CDC), which includes the Agency for Toxic Substances and Disease Registry, is recognized as a lead federal agency for protecting the health and safety of people at home and abroad, providing credible information to enhance health decisions, and promoting health through strong partnerships.

Research is the foundation of CDC's success and enables CDC to develop, improve and disseminate evidence-based interventions, programs, and decision support to improve health (Figure I.1). The primary focus of CD C's research is to fill gaps in knowledge necessary to accomplish the agency's Health Protection Goals (Chapter II), but it must also be CDC's core public health mission and be responsive to new opportunities, threats, and future health needs. A recent analysis indicates that the evidence base to support much of public health practice has not kept pace with the growing requirements, and much more research is needed to demonstrate the best methods that support effective public health practices[1]. The new Advancing the Nation's Health: A Guide to Public Health Research Needs, 2006–2015[2] (also referred to as the Research Guide) will serve as a critical resource for research areas that should be addressed during the next decade by CDC and its partners in response to current and future public health needs and events.

---

[1] Thacker SB, Ikeda RM, Gieseker, KE, Mendelsohn, AB, Saydah, SH, Curry, CW, Yuan, JW. The evidence base for public health: informing policy at the Centers for Disease Control and Prevention. Am J Prev Med 2005; 28(3):227-233.

[2] Advancing the Nation's Health: A Guide to Public Health Research Needs, 2006–2015 was formerly titled Health Protection Research Guide, 2006–2015 during the public comment period (See Chapter I Section E).

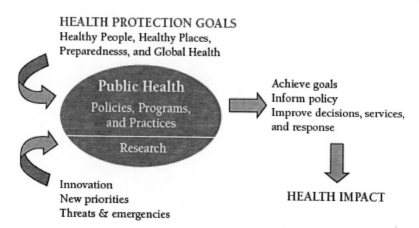

Figure I.1. CDC's Integration of Goals, Research and Programs.

By health protection research, we mean research that supports health promotion, prevention of injury, disability, and diseases, and preparedness activities[3]. Many federal agencies have a distinguished and long track record of support of health-related research in their specific focus areas, and many agencies are guided by their own research agendas. What distinguishes the Research Guide from other health-related research agendas is that it is the first-ever comprehensive, long-term, national resource spanning all areas of health protection that will allow us to shine a spotlight on public health as never before. As such, the Research Guide is intended to build on and complement the work of other federal agencies and their research agendas.

The Research Guide will also serve as an essential resource for defining a more focused CDC health protection research agenda of research priorities aligned with the Health Protection Goals developed by CDC. In addition, portions of the Research Guide will be used to inform research initiatives that address other critical public health needs and research priorities of other agencies. CDC has developed the Research Guide with extensive input from its staff and a wide range of partners and stakeholders, including external researchers, other federal agencies, state and local health departments, professional associations, universities, non-governmental organizations, business and worker organizations, community groups, American Indian and Alaska Native governments, tribal leaders and organizations, and the public-at-large.

---

[3] Research is defined as a systematic investigation, including research development, testing and evaluation, designed to develop or contribute to generalizable knowledge (Department of Health and Human Services. Code of Federal Regulations: Title 45, Subpart A, Section 46.102. Available at http://www.hhs.gov/ohrp/ humansubjects/guidance/45cfr46.htm#46. 102).

B. RATIONALE

The challenges of our increasingly complex and interdependent world require new approaches to generating and disseminating the knowledge and innovations needed to promote well-being and improve health. In response to these challenges, CDC recently underwent an agency-wide strategic planning process called the Futures Initiative. Through this initiative, CDC has renewed its commitment to improving and protecting health around the globe by moving to a new organizational structure (Appendix I), which was designed to enhance coordination and collaboration within the agency, and with its partners and the public. The Futures Initiative has prompted the agency to closely examine the research it conducts. CDC has also recently developed six strategic imperatives that are intimately linked to research (Table I.1).

Table I.1. CDC Strategic Imperatives and Their Relationship to Research*

Health Impact Focus: Align CDC's staff, strategies, goals, investments, and performance to maximize impact on the population's health and safety. Research provides knowledge of how to focus on areas of greatest health impact.
Customer-centricity: Market what people want and need to choose health. Research reveals new ways to support customer needs and priorities.
Public Health Research: Create and disseminate the knowledge and innovations people need to protect their health now and in the future. Research supports the scientific foundation of public health policies, programs, and practices.
Leadership: Leverage CDC's unique expertise, partnerships, and networks to improve the health system. Research provides the evidence-base for informed decision- making and therefore is an essential building block for a strong national public health system.
Global Health Impact: Extend CDC's knowledge and tools to promote health protection around the world. Research develops new methods and tools tailored to the varied needs of diverse populations around the world.
Accountability: Sustain people's trust and confidence by making the most efficient and effective use of their investment in CDC. Research resources are focused on the most pressing public health problems to ensure that these resources are used wisely.

*The strategic imperatives are in regular typeface. The relationships of research to their
    respective strategic imperatives are italicized.

The Research Guide can play an important role in defining the research necessary to ensure that each of these six strategic imperatives is fulfilled. Research is also critical to achieving the Health Protection Goals (Chapter II) which were developed by CDC and designed to focus public health activities towards accelerating the achievement of a greater positive impact on people's health. To reach its strategic imperatives and the Health Protection Goals, CDC is collaborating with academic researchers, federal agencies, state and local health departments, education agencies; national, international, professional, and community- based organizations; and the general public. Through these strategic alliances and research funding mechanisms, promising research can more rapidly be translated and disseminated through practical cost-effective policies, programs, and practices that improve health.

C. PURPOSE

Advancing the Nation's Health: A Guide to Public Health Research Needs, 2006–2015 will serve the following purposes:

1. *Identify knowledge gaps that must be considered as CDC develops its Health Protection Goals Action Plans.* CDC has developed Health Protection Goals (Chapter II) and is developing Goals Action Plans for achieving significant impacts on the health of the U.S. population and the world.

2. *Describe the range of research most needed to provide critical evidence for the improvement of existing and the establishment of new public health programs and interventions.* A distinguishing and essential feature of CDC research is that it fills critical gaps necessary to improve public health programs, services, and response.

3. *Improve the effectiveness of a broad range of public health disciplines through supporting innovative, cross-cutting, interdisciplinary and foundational research.* This research may cut across several public health fields and thus have a profound impact on the ability to protect and improve public health.

4. *Serve as a platform for discussions with federal partners about opportunities to collaborate in addressing the most pressing current and future public health problems.* Many public health needs can best be met through a coordinated research strategy involving multiple federal agencies, which together can leverage their unique strengths and resources to more effectively solve national and global health problems Provide a basis for discussions with state and local partners about identifying opportunities for collaboration to better address health needs across the United States. CDC will identify research gaps, fulfill research

needs, and communicate research findings by collaborating with state and local entities that carry out the work of public health.

5. *Promote opportunities to partner with academic institutions, professional associations, international agencies, tribal organizations, businesses, worker organizations, and non-profit and community-based organizations to address institutional and community research needs.* Working with a wide range of partners to address public health can ensure that research findings are beneficial, practical and tailored to all groups of people and their communities.

6. *Plan for and promote public health research needs.* To be successful, effective communication and promotion of research needs and priorities are needed to improve health and public health practice.

D. SCOPE AND USE OF THE RESEARCH GUIDE

The Research Guide's priority research needs are expressed at a broad level to represent the full range of research that may be conducted within an extended timeframe. This breadth also permits inclusion of new priorities that may arise in the future from emerging threats and needs. The Research Guide is comprehensive because it engages and provides guidance to the broad research community within and outside of CDC.

The Research Guide is heavily focused on producing new knowledge that will support CDC's Health Protection Goals. The Research Guide also includes some types of crucial research that are indirectly related to these Goals, including innovative research, and cross-cutting research (e.g., data science, social determinants of health, and public health law), which serve as a foundation for many public health disciplines.

The research contained in the Research Guide is also aligned with Healthy People 2010 (HP 2010) and serves as tool for identifying critical public health research needed to support HP 2010 objectives [1]. Appendix II shows the research categories and themes that address research gaps associated with the HP 2010 Leading Health Indicators and related objectives. The Leading Health Indicators were developed under HP 2010 to measure the health of the nation over the next ten years. These indicators are linked to one or more objectives that reflect the major health concerns in the United States at the start of the 21st century. The Research Guide will be applicable to both intramural and extramural research sponsored by CDC and other agencies. CDC Coordinating Centers, Coordinating Offices, and individual Centers, Institute, and Offices will continue to have primary responsibility for developing and overseeing extramural and intramural research programs that fall under the agency-wide Research Guide, the Goals Action Plans, and categorical research agendas.

In the interest of maximizing improved health impacts, CDC will collaborate with many partners to build and strengthen networks of research to improve public health interventions. Through a coordinated approach, critical knowledge gaps outlined in the Research Guide can be filled by CDC and multiple partners. The Research Guide provides an extensive landscape of research needs in many different areas of public health.

Effective networks focused on specific research areas or themes are becoming a critically important strategy. A "one-government" approach has become a reality, as well as using multidisciplinary approaches that reach across and among agencies and partners to solve complex public health problems. CDC is committed to effective partnerships and using them to advance public health.

Development of Short-Term Research Priorities

CDC's new Goals implementation process will influence the future allocation of resources for all of CDC's activities, including research. The Research Guide will be a critical resource for the development of a more focused CDC health protection research agenda of research priorities that will emerge as the Goals Action Plans define critical knowledge gaps (Chapter II). The initial research initiatives identified by the Goals Action Plans will become part of the ongoing research agenda for CDC. To facilitate this, all of the Research Guide's research themes, except for cross-cutting research, are mapped to the Overarching Health Protection Goals (Table II.1).

E. RESEARCH GUIDE DEVELOPMENT PROCESS

Development of Early Drafts

In 2001–2004, CDC convened three large workgroups comprised of CDC staff that were tasked with recommending a process for developing an agency-wide research agenda. In January 2005, CDC established six workgroups and a core team comprised of representatives from state and local health departments, academic institutions, advocacy groups, partners, and Centers, Institute, and Offices within CDC (Appendix III). The six workgroups and core team each included a member external to CDC who was selected for his or her expertise in health disparities research. The workgroups and core team were tasked with implementing the recommended process to develop the Research Guide (see Table I.2 for dates of key milestones) The workgroups and core team were advised by a group of senior external advisors who comprised the Research Agenda Steering Subworkgroup, of the Workgroup on Goals and Research Agenda, of the Advisory Committee to the Director, CDC (Appendix III). The workgroups used standardized criteria and consulted CDC staff and external partners to develop a draft Starter List of Research Priorities that was used to stimulate discussion on the potential research themes to be included in the Research Guide.

Table I.2. Timeline of Research Guide Development Key Milestones

Milestone	Date
Develop and propose research agenda development process	2001-2004
Launch workgroups and hold orientation	January 10, 2005
Gather CDC staff input	January-April, 2005
Create Starter List of Research Priorities	February 25, 2005
Hold four public participation meetings	March 8-31, 2005
Hold first federal partners meeting	March 9, 2005
Produce public comment draft	May-November, 2005
Sponsor 60-day public comment period	November 18, 2005 –January 15, 2006
Vet and finalize Research Guide	Summer – Fall 2006
Update the Research Guide	Periodically as needed

Prioritization Criteria for Research Theme Candidates

Each research theme candidate was evaluated and ranked on the following four standardized criteria.

1. Relevance to CDC's mission and the Health Protection Goals
2. Importance of the problem and public health need being addressed
3. Relevance to reducing health disparities

4.   Potential for broad impact on multiple diseases or risk factors

Other factors also were considered for determining which research themes were appropriate for inclusion in the draft Starter List of Research Priorities. Research themes addressing areas CDC deemed as needing immediate attention (e.g., obesity, adolescent health, preparedness, and influenza) were prioritized for inclusion, as well as research ideas that were likely to have the greatest short-term impact on public health (e.g., research on the efficacy, effectiveness, and dissemination of public health interventions).

To ensure that cross-cutting research achieved prominence in the Research Guide, additional priority was given to research themes addressing public-health-workforce capacity and training, the social determinants of health, and innovation and to those themes that could significantly impact many different sectors (e.g., public health systems, health-care-delivery systems, educational institutions, and private industry) The original 129 research theme candidates that ranked highest by the workgroups were included on the Starter List of Research Priorities. These themes were classified to one of the following seven categories.

1.   Infectious diseases
2.   Community preparedness and response
3.   Health promotion
4.   Environmental and occupational health and injury prevention
5.   Global health
6.   Health information and services
7.   Cross-cutting research

Initial Public Participation

Once the draft Starter List was compiled, CDC gathered public input to prepare for creating the initial draft of the Research Guide. Public participation meetings were held during March 2005 in Arlington, Virginia; Atlanta, Georgia; Seattle, Washington; and Columbus, Ohio. At these four meetings, CDC received input from a diverse group of 450 researchers, representatives of partner organizations, and the public-at-large. The draft Starter List of Research Priorities for the Research Guide was also made available to the public for comment through an Internet website. Additionally, all CDC employees were invited to submit comments. In March 2005, CDC also held its first meeting with other federal agencies, largely those within the U.S. Department of Health and Human

Services, to discuss how to build and enhance federal networks to promote the use of the Research Guide. Planning for additional meetings with federal partners to continue and expand these efforts is underway.

Each of the comments and recommendations received through the above-mentioned venues was considered by members of the workgroups that developed the Research Guide. Modifications and additions were made to the Research Guide on the basis of this feedback, resulting in the draft version of 131 research themes released on November 18, 2005. The draft version was reviewed and approved by CDC leadership and senior scientific staff, as well as key external advisors.

Public Comment Period

CDC accepted public comments on the draft Research Guide for 60 days (from November 18, 2005 through January 15, 2006). Public comments were accepted in three ways: a) the Internet (via a public comment website), b) e-mail (via the ResearchGuide@cdc. gov mailbox), and c) letters (via the U.S. Postal Service). CDC received 491 comments through the website, 6,726 comments through e-mail, and 8 sets of comments by U.S. postal service, although several of the latter comments were also submitted via the web or e-mail. These comments were then broken down into individual, discrete ideas and distributed to the appropriate workgroups for consideration. Public feedback was requested in the following areas:

- Scope and use of the Research Guide (including whether it identified the areas of health protection research most needed to be addressed within the next decade)
- Relevance and level of specificity of the proposed research topics
- Additions, deletions, or modifications to the proposed research topics
- The *Research Guide* development process
- Other improvements to the *Research Guide*

Finalizing the Research Guide

After the public comment period, the Research Guide development workgroups reconvened to consider and address comments received from the public to produce a final version of the Research Guide. One new research category, seven new research themes, and a table showing the relationship between Healthy People 2010 Leading Health Indicators [1] and the research themes (Appendix II) were added to the Research Guide, along with many other revisions, as a result of the public comments. The final version of the Research Guide contains 138 research themes. The Research Guide was reviewed and vetted by members of CDC leadership, CDC senior scientific staff, and CDC's external advisory committees before being published.

Updating the Research Guide

Because the Research Guide is meant to be a dynamic document, it will be revised and updated as needed in response to unexpected or unanticipated needs, new findings, or new goals. Greater detail regarding the Research Guide development process and future updates can be found at CDC's website http://www.cdc.gov/od/science/PHResearch/cdcra/index.htm.

F. ORGANIZATION OF THE *RESEARCH GUIDE*

The Research Guide is organized into three tiers (Figure I.2). The broadest tier consists of the seven major research areas that are represented by Chapters III through IX, respectively.

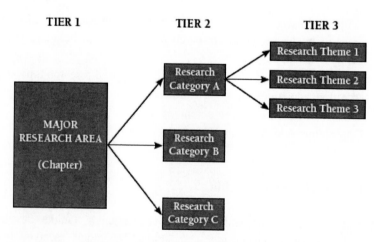

Figure I.2. Organization of the Research Guide.

The major research areas represent an evolution of the seven broad research categories contained in the draft Starter List of Research Priorities. (The relationship between the Starter List categories and the corresponding Research Guide chapters is shown in Table I.3.)

Table I.3. Relationship Between Starter List of Research Priorities Categories and Research Guide Chapters

Starter List Category Title	Research Guide Chapter Title
Infectious Diseases	Prevent and Control Infectious Diseases
Community Preparedness and Response	Promote Preparedness to Protect Health
Health Promotion	Promote Health to Reduce Chronic Diseases and Disability
Environmental and Occupational Health and Injury Prevention	Create Safer and Healthier Places
Global Health	Work Together to Build a Healthy World
Health Information and Services	Manage and Market Health Information
Cross-Cutting Research*	Promote Cross-Cutting Public Health Research*

* Because cross-cutting research priorities are intended to infuse all areas of the Research Guide, some overlap exists among the cross-cutting research described in Chapter IX and the other more discipline-specific research categories and themes that appear elsewhere in the Research Guide.

Each chapter addressing a major research area contains three to eight specific research categories that further describe the major research area. Each research category contains one to ten more specific research themes that further define the research within that research category. Each research theme presents examples of priority research designed to specify research at a level that could be described in a collection of related Funding Opportunity Announcements.

Chapter II

RELATIONSHIP OF RESEARCH THEMES TO THE CDC HEALTH PROTECTION GOALS

A. CDC HEALTH PROTECTION GOALS

CDC is committed to achieving measurable and significant improvements in the health of all persons living in the United States. To further its efforts, the agency has defined specific Health Protection Goals, which will help CDC prioritize and focus its work and investments and enable the agency to measure progress for all of its activities, including research. CDC has developed four overarching Health Protection Goals: healthy people, healthy places, preparedness, and global health (Table II.1). Each overarching Goal is associated with several more specific Goals. (See http://www.cdc.gov/about/goals for additional information about the Goals).

The CDC healthy people Goals were created to address the unique health protection requirements that vary by stage of life. To better meet each group's unique health concerns and risks, Goals have been created for each of the five life stages: infants and toddlers, children, adolescents, adults, and older adults and seniors.

The CDC healthy places Goals focus on the environments in which people live, work, learn, and play. With a focus on communities including places of worship, homes, health-care settings, schools, institutions, workplaces, and the places we travel and use for recreation, the activities associated with the healthy places Goals will help ensure a holistic approach to achieving health across the lifespan in all environments, from birth through retirement and beyond. By improving the quality and safety of our environments, CDC and its partners will continue protecting and improving the health of all persons living in the United States in all aspects of their daily lives.

Table II.1. Overarching Health Protection Goals

1. Healthy People in Every Stage of Life
All people, especially those at higher risk due to health disparities, will achieve their optimal lifespan with the best possible quality of health in every stage of life.
2. Healthy People in Healthy Places
The places where people live, work, learn, and play will protect and promote their health and safety, especially those at greater risk of health disparities.
3. People Prepared for Emerging Health Threats
People in all communities will be protected from infectious, occupational, environmental, and terrorist threats.
4. Healthy People in a Healthy World
People around the world will live safer, healthier, and longer lives through health promotion, health protection, and health diplomacy.

The CDC preparedness Goals focus on improving capacity to respond to health threats, including those associated with naturally and intentionally introduced biological, environmental, chemical, radiologic, bombing, and other explosive events. The preparedness Goals focus not only on improving public health capacity, but also on timely and accurate identification and reporting of health threats; dissemination of information regarding those threats; provision of countermeasures and health guidance to those affected; reduction in the time needed to restore safety; improvement in long-term follow-up after an event occurs; and reduction in response time following public health threats.

The CDC global Goals seek to promote health around the world through sharing knowledge, tools, and other resources; to protect against health threats at home and abroad through improved transnational prevention, detection, and response networks; and to promote and protect global health through effective health diplomacy.

To set priorities and measure progress, CDC is currently developing Goals Action Plans, which will specify prioritized objectives and activities, including research, for the CDC Health Protection Goals.

B. RELATIONSHIP OF RESEARCH THEMES TO THE CDC HEALTH PROTECTION GOALS

CDC strives to ensure that research dollars are invested initiatives that will result in the greatest health impact. Good management of research involves identifying research priorities, translating these priorities into a balanced portfolio of intramural and extramural research, evaluating research programs and outcomes, coordinating closely with federal partners, and incorporating lessons learned from evaluation into future research priorities. These are the principles of CDC's Goals implementation process. The research proposed in Advancing the Nation's Health: A Guide to Public Health Research Needs, 2006–2015 is designed to be integrated with this process and the CDC Health Protection Goals.

To illustrate this alignment, the research themes included in the Research Guide have been mapped to the overarching Health Protection Goals (Appendix IV). Because the research themes specify long-term research priorities, they are relatively broad (See Scope and Use of the Research Guide, Chapter I). Thus, some research themes map to more than one overarching Goal. In addition, the Research Guide contains some research themes that are not directly related to specific Health Protection Goals, but support a range of goals. These themes are classified as "cross-cutting" research in Appendix IV.

Short-term public health research priorities will be identified in the Goals Action Plans that are currently under development. The comprehensive Research Guide will serve as an essential source for existing research needs in the development of short-term priorities as part of the Goals Action Plans. If the Goals Action Plans identify additional critical science gaps and research needs that are not contained in the Research Guide, the latter will be updated to reflect these needs.

Chapter III

PREVENT AND CONTROL INFECTIOUS DISEASES

Infectious diseases are an ever-present threat to the well-being of all people around the globe, regardless of nationality, age, sex, lifestyle, race, ethnic background, disability, sexual behavior and identity, and socioeconomic status. Although all populations are at risk for certain infectious diseases, the degree to which some populations are impacted by them can vary. Persons who are very young or old, patients who are hospitalized, and persons whose immune systems are compromised are particularly at risk for the negative health effects associated with infectious diseases. The individual characteristics of each person within a population also can make some persons more vulnerable to infectious diseases than others.

Despite modern advances, many infectious diseases continue to inflict unnecessary suffering and death. Diseases once thought to be controlled can re-emerge in new forms that do not respond to antimicrobial drugs or other existing interventions; novel infectious diseases for which no interventions have been developed also continue to emerge, challenging the ability of public health and other responders to stop them from spreading.

Public health research is crucial to ensuring adequate public health response to known and emerging infectious agents, the prevention of infection-associated chronic illnesses, and increased understanding of the underlying reasons for infectious-disease-associated health disparities. Infectious disease research will help protect people, especially those at high risk, from infectious threats, including those that occur naturally and those used in intentional acts of terror. In addition, infectious disease research can lead to a) the development of new and increased use of existing interventions known to prevent biological agents and naturally occurring infectious diseases; b) more timely and accurate diagnostic tests and communications regarding public health threats; c) a quicker public health response in identifying the causes, risk factors, and appropriate interventions for infectious diseases; and d) the prompt provision of countermeasures and health guidance to persons who are most affected by emerging and re-emerging infectious diseases. Research categories in this chapter include:

A. Emerging and Re-emerging Infectious Diseases
B. Pandemic and Seasonal Influenza
C. Infectious Disease Surveillance and Response
D. Vaccines and Immunization Programs
E. Behavioral, Social, and Economic Research in Infectious Diseases
F. Host-Agent Interactions
G. Special Populations and Infectious Diseases

A. EMERGING AND RE-EMERGING INFECTIOUS DISEASES

Advances in science, technology, and medicine enable the treatment and prevention of many diseases that historically have caused high rates of morbidity and mortality around the world. However, other advances (e.g., globalization of travel and trade) have created new opportunities for microbes to emerge and spread. Resource problems, trade concerns, and low prioritization of infectious diseases, which contribute to inadequate reporting and control of emerging threats in some countries, also have begun to contribute to the excess spread of disease. In addition, health threats that were inconceivable only a few decades ago are now being recognized; people around the globe are starting to face infectious disease threats resulting from terrorist acts involving the deliberate release of pathogenic microbes with intent to cause harm.

> Infectious diseases are one of the oldest threats to human health. Improving the detection, prevention, and control of these diseases can better prepare and protect our nation from new and re-emerging disease threats.

Infectious microbes have a remarkable ability to evolve and adapt to new situations, sometimes becoming more virulent or developing resistance to commonly used drugs [2]; microbial evolution can render vaccines and therapeutic drugs ineffective. As a result, unique public health challenges and scientific questions have been associated with the recent emergence and re-emergence of infectious diseases around the globe, including human immunodeficiency virus (HIV), tuberculosis (TB) (caused by Mycobacterium tuberculosis complex), malaria (caused by Plasmodium species), hepatitis C, West Nile virus, mad cow disease (variant Creutzfeldt-Jakob disease/bovine spongiform encephalopathy), severe acute respiratory syndrome (SARS), and community-associated methicillin-resistant Staphylococcus aureus. Although these specific diseases are not mentioned by name in the following sections of this chapter, each proposed research initiative can be applied to these and other specific infectious disease threats to public health.

A significant risk factor for disease emergence in the years ahead is likely to be population growth and urbanization [3], which will lead to the creation of megacities throughout the world. These population increases will be associated with ecological changes that allow increased contact among humans, animals, insects, and animalborne and insectborne pathogens. Improving research that focuses on the prevention and control of emerging and re-emerging infectious diseases will help ensure that people in all communities are protected from infectious threats.

1. Antimicrobial Resistance

Develop strategies to improve appropriate antimicrobial use, conduct clinical trials to validate use of existing drugs, and develop improved tools for conducting epidemiologic and microbiologic studies and surveillance of drug resistance.

Examples of Priority Research: Identify methods to preserve the effectiveness of antimicrobial agents. Develop, implement, and evaluate the impact of control policies and interventions to improve the use of antimicrobial agents in human and veterinary medicine. Develop and evaluate methods for preventing the spread of antimicrobial resistant pathogens (particularly those that are multidrug-resistant) in diverse settings (e.g., the community, health-care facilities, and farms). Conduct clinical trials of existing drugs for which no valid trial data exist regarding efficacy for antimicrobial-resistant infections (i.e., drugs used to prevent and treat certain infections for which efficacy remains unproven). Conduct clinical trials, ideally in collaboration with industry, to determine the efficacy and safety of new drugs and biologics that can be used to treat and control drug-resistant infections. Develop methods of surveillance that employ standardized and accurate laboratory methods and that link microbiology, drug use, and health-outcomes data, allowing for estimation of the cost-effectiveness of various empiric therapies for infections both in and outside of health-care facilities. Develop tools for the rapid detection of the genetic determinants of antimicrobial resistance in microbes located within environmental and clinical samples. Determine the optimal methods of preventing the spread of drug-resistant infections in health-care facilities and other settings or cohorts where outbreaks have been described, and evaluate the efficacy of relevant vaccines to decrease infections.

2. Environmental Microbiology of Bioterrorism-Related and Other Pathogens

Develop methods to establish the presence of microbial pathogens in the environment, estimate the risk of infection to human populations, and develop strategies to implement proven risk-reduction strategies for infectious disease transmission in contaminated environments.

Examples of Priority Research: Develop effective sampling strategies for and sensitive methods of obtaining specimens and identifying infectious agents in the environment. Develop sensitive and specific methods to test for a range of priority agents (including Category A agents, microbial agents in Categories B & C [4] other microbial agents, and toxins produced by microbial agents) in a variety of environmental media (e.g., water, air, surface, and bulk materials). Develop techniques to classify organisms with maximal specificity using genomic, proteomic, and atomic characteristics. Conduct studies on the efficiency of dose and exposure routes for delivering infectious pathogens and/or toxins through different environmental media (e.g., water, food, air, surface, and bulk materials). Conduct studies on the durability of infectious pathogens in environmental media (including areas involved in personal decontamination, such as skin, hair, and clothes) and under different conditions (e.g., temperature, humidity, pH, and light level). Develop technologies and procedures for reducing human exposure and removing or inactivating pathogens in contaminated environments, including naturally occurring bioterrorism agents.

3. Emerging Infections and New Prevention Technologies

Improve rapid histopathologic and immunopathologic detection of infectious agents; develop methods for the rapid detection of known families of bacteria, viruses, parasites, and fungi; improve prevention of zoonotic diseases; detect illness among donors and recipients of solid organ, tissue, and other cellular transplants; and improve health-care preparedness.

Examples of Priority Research: Develop and assess new technologies for the prevention of infectious diseases. Develop new methods of histopathologic and immunopathologic detection of infectious agents. Develop surveillance and laboratory tools to identify new and emerging pathogens. Conduct research on human, microbial, and environmental factors that contribute to the emergence of infectious agents. Develop a comprehensive set of microbial DNA microarrays that contains phylogenetic sequences from all known families of bacteria, viruses, parasites, and fungi and can be used to rapidly determine the cause of an outbreak of unknown etiology. Develop methods to prevent transmission of infectious diseases through solid organ, tissue, and other cellular transplants. Develop strategies for monitoring infection in donors and recipients of solid organ, tissue, and other cellular transplants. Develop animal models for transmission of infectious agents to improve understanding of the pathogenesis of emerging infections transmitted through biologic tissue. Establish best practices for effective responses to emerging or intentional infectious agent exposure events in health-care settings to prevent transmission to other patients, health-care workers, and the community. Develop and assess the impact of effective criteria and tools for screening patients, visitors, and health-care workers who are or may be infected with epidemic strains of infectious agents; examine the use of these criteria and tools during various stages of disease activity at the local level. Measure the effectiveness of infection-control interventions on limiting dissemination of respiratory infections in various health-care settings. Assess the effectiveness of syndromic surveillance, triage, and other early-detection systems in identifying and isolating patients with pandemic influenza or other emerging infectious diseases in health-care settings. Evaluate methods of personal protection (e.g., use of microbicides, disinfectants, and dental decay prevention technologies) for averting various infections.

4. Health-Care-Associated Infections and Patient Safety

Create novel strategies for preventing health-care-associated infections, including those occurring among recipients of solid organ, tissue, and other cellular transplants.

Examples of Priority Research: Evaluate the use of enhanced electronic reporting systems and surrogate markers to detect and prevent adverse events associated with invasive devices, invasive procedures, biologic products, and medication. Develop improved methods for identifying and preventing health-care-associated infections and for determining relatedness of pathogen strains. Develop strategies to prevent unsafe injections in health-care settings. Develop methods to decrease errors associated with the identification and antimicrobial susceptibility testing of health-care-associated infections. Develop methods for promptly and accurately communicating and integrating information from all points in the process of procuring, donating, and transfusing or transplanting solid organ, tissue, and other cellular transplants (e.g., blood collection centers, organ and tissue procurement organizations, public health authorities, and adverse-event surveillance systems). Conduct studies aimed at decreasing infections transmitted from cadaveric organs and allograft tissues through a) improving donor screening; b) tracking allograft tissues; c) standardizing methods for maintaining quality in the processing and storage of tissues; and d) detecting allograft-associated infections. Determine the efficacy of masks and respirators in preventing the transmission of various microorganisms between patients and health-care workers. Determine the contribution of fit-testing in ensuring that respirators provide protection to health-care workers and emergency responders. Evaluate the most effective personal respiratory protection programs, and quantify the relative importance of respirators in preventing infectious disease transmission compared with other control measures. Determine institutional, facility, or health-care provider factors that facilitate or impede infection-control practices.

5. Zoonotic and Vectorborne Diseases (ZVBDs)

Develop new methods of detection, control, and prevention of existing and emerging ZVBDs in humans and animal populations.

Examples of Priority Research: Characterize newly identified pathogens in domestic and wildlife species of animals that can infect humans. Develop reliable laboratory assays for ZVBDs. Determine new targets and tools for diagnostic applications, and develop new and improved diagnostic assays. Conduct studies to improve diagnostics for the early detection and surveillance of ZVBDs in animals that precede human infection. Develop surveillance methods for the emergence of ZVBDs globally. Create models to assess factors that predict the risk of, contribute to the spread of, and may limit efforts to control ZVBDs by integrating climatic, environmental, veterinary, ecologic, entomologic, and epidemiologic data. Conduct research on the human host, microbial, environmental, and behavioral risk factors that contribute to the emergence of ZVBDs. Identify intervention and control measures for ZVBDs. Conduct surveillance to assess the impact of intervention and control measures on ZVBDs. Collaborate with animal health experts to develop strategies from wildlife, agricultural, aquatic, companion animal, and captive exotic species sectors to address ZVBDs. Determine the actual imported zoonoses risks at U.S. ports-of-entry via animals and animal products. Establish standards and procedures for conducting surveillance for and managing pesticide resistance in vector mosquitoes. Develop standardized pesticide resistance test kits for use by local health agencies.

B. PANDEMIC AND SEASONAL INFLUENZA

Both pandemic (i.e., a worldwide epidemic resulting from efficient transmission of a novel influenza A virus to which there is little or no pre-existing human immunity) and interpandemic (commonly referred to as "seasonal") influenza pose tremendous threats to the health and life of persons living in the United States. In particular, pandemic influenza has the potential to overwhelm the U.S. health-care delivery system's capacity to adequately respond to a sudden and enormous demand. A seasonal epidemic occurs predictably every winter in the United States, resulting in about 36,000 deaths and 200,000 hospitalizations, mostly among the elderly [5]. By contrast, an influenza pandemic has the capacity for catastrophic global impact. Such a disease event would likely cause extensive morbidity and mortality, social disruption, and economic loss [6]. The H5N 1 strain of avian influenza is an example of a virus that could mutate into a new strain capable of causing the next influenza pandemic. Although both types of influenza have the potential to impact all populations, some groups (e.g., young children, the elderly, pregnant women, persons with certain pre-existing health conditions, and those whose immune systems are compromised) might be more susceptible to infection, severe disease, and death. The health of certain people (e.g., those who are economically disadvantaged, are elderly, or have disabilities) may be disproportionately impacted because of limited access to already scarce, life-saving health-care resources. Steps can be taken now to substantially reduce the health, social, and economic impacts that both pandemic and seasonal influenza epidemics will have on the general population and persons at high risk. Research in influenza prevention and control, including elucidating the optimal means of antiviral and vaccine distribution, will prepare the nation to better protect people in all communities from this infectious threat. Although this section is focused on influenza, the principles enumerated in the following paragraphs can be applied to other diseases with epidemic and/or pandemic potential.

1. Pandemic and Seasonal Influenza

Determine the pathogenesis, transmission, and immune response of highly pathogenic avian and other influenza viruses with pandemic potential to develop improved preventive and therapeutic measures.

Examples of Priority Research: Examine the pathogenesis and transmission of highly pathogenic avian influenza viruses (HPAIs) in mammalian models. Improve reliability of diagnostic methods for influenza, focusing on strains with pandemic potential. Assess the degree of existing immunity to HPAIs and determine the risk factors (i.e., viral, host, and environmental) for transmission of HPAIs from birds to humans and from person to person. Examine the evolution and antigenic variation of HPAI. Develop vaccine candidates for potential pandemic viruses using traditional and reverse-genetics approaches. Evaluate the effectiveness of various technologies (e.g., egg- and cell-based technologies and alternative delivery and antigen-sparing technologies) on vaccine uptake and effectiveness in general populations and among high risk groups. Evaluate the immune response to HPAI and vaccines in high-risk populations, especially persons who are immunocompromised, the elderly, children, and pregnant women. Evaluate immunologic correlates of protection among persons with severe illness in various stages of life. Model the effectiveness of vaccination strategies in cases of limited vaccine availability. Assess the utility of immunization registries and vaccine tracking systems for pandemic and non-pandemic influenza vaccines. Develop and test materials and methods to rapidly monitor vaccine-associated adverse events during a pandemic. Evaluate the effectiveness and safety of antiviral drugs against influenza strains with pandemic potential, and determine the field effectiveness for containing localized outbreaks using a combination of enhanced surveillance, rapid laboratory diagnostics, community containment, and high penetration of prophylactic viral medications. Improve diagnostic methods for influenza, and develop streamlined screening approaches for monitoring the emergence of drug-resistant strains during a pandemic. Assess the impact of age and other host-specific factors on hospitalizations and death. Evaluate the effectiveness of immunization and therapeutic approaches in high-risk populations. Evaluate and field- test communication messages, and determine the efficacy of "low-tech" intervention measures (e.g., hand hygiene and face masks). Investigate the most effective ways to relate uncertainty to the public. Model and evaluate the feasibility and effectiveness of large-scale disease containment strategies (e.g., social distancing, the cancellation of mass gatherings, "snow days," group quarantine, and area quarantine). Evaluate the potential impact of universal influenza vaccination recommendations on vaccine uptake and supply. Conduct modeling studies to assess health-care utilization and surge capacity (e.g., beds, staff, antimicrobial drugs, and ventilators) for different scenarios.

C. INFECTIOUS DISEASE SURVEILLANCE AND RESPONSE

Rapid, inexpensive, sensitive, and specific etiologic diagnostic tests that can enhance the treatment and control of infectious diseases are often lacking. Additional prevention research relevant to virtually all infectious diseases is needed in light of diminishing new vaccine and antimicrobial development [7]. Many infectious diseases disproportionately affect certain populations (2) (e.g., minorities, children, adolescents, men who have sex with men, and persons who are economically disadvantaged and foreign born), resulting in health disparities. Evaluating various prevention methods to identify which methods are most effective for high-risk populations and in settings where rates of specific infections are higher would help reduce these disparities. Development of new diagnostic methods would prevent many infections, ensure appropriate therapy, and inform the design of control strategies (e.g., vaccination). New methods of prevention and control are likely to be applicable to multiple infectious diseases. Accurate and rapid diagnostic testing is crucial for determining appropriate therapy, controlling institutional and community outbreaks, describing disease burden, and responding effectively to a pandemic or potential bioterrorism event. Research in the prevention of infectious diseases supports the basic public health objective of ensuring that people in all communities are protected from infectious disease-related threats.

1. Infectious Disease Diagnostic Methods

Develop rapid, inexpensive, sensitive, and specific etiologic diagnostic tests for a variety of infectious disease agents.

Examples of Priority Research: Develop improved methods of diagnosing respiratory illnesses of public health importance to expedite implementation of appropriate control measures. Develop accurate, rapid, and affordable diagnostic tests capable of detecting multiple agents (e.g., viral, bacterial, and fungal) from a single sample which will enable clinicians and public health workers to deliver treatment, prevention, and control strategies more rapidly. Develop reliable laboratory assays for viral and rickettsial diseases of public health importance. Expand the capacity for diagnosis and molecular typing to discover new targets and tools for diagnostic applications. Develop molecular and immunologic tools for the surveillance and diagnosis of microbes that cause known or potential vaccine-preventable diseases. Develop diagnostics for viral, bacterial, and fungal pathogens that are a threat to solid organ, tissue, and other cellular transplant/transfusion recipients. Develop strategies to increase effective use of diagnostics in clinical practice, and determine how the use of diagnostics and pathologic examination can improve infectious disease prevention and control.

2. Infectious Disease Surveillance and Response

Improve infectious disease surveillance and the timeliness of response for pathogens that affect U.S. and international populations.

Examples of Priority Research: Develop more advanced, practical disease surveillance systems by combining laboratory subtyping (i.e., "fingerprinting") with epidemiologic data. Develop more efficient tools for outbreak detection, investigation, and reporting that integrate clinical, epidemiologic, and laboratory data obtained from various sources (e.g., health-care providers and public health institutions at the national, state, and local levels). Using geographic information systems (GIS), evaluate the usefulness of syndromic surveillance and outbreak mapping, and determine the way these activities impact disease detection and control. Enhance the recognition of novel, previously unidentified pathogens through examination of the genes they carry; improve capabilities for detecting, characterizing, and fingerprinting known pathogens rapidly and with high accuracy using cutting-edge methods (e.g., DNA and protein sequencing and mass spectrometry); and learn more about the prevalence and ecology of known and newly identified pathogens. Conduct research to optimize the use of electronic health records for public health surveillance purposes in the context of privacy protection (e.g., the Health Insurance Portability and Accountability Act [8]). Evaluate the effectiveness of new sources of real-time or near real-time disease detection and the monitoring of healthcare delivery practices. Evaluate proposed and existing information systems and linkages between them for their usefulness in the detection of infectious disease epidemics; examine their potential for detection of the major biothreat agents, their ability to monitor the spread of epidemics, and their cost-effectiveness. Enhance methods used by national and international networks that identify the spread of pathogens across borders. Develop new tools and strategies for controlling neglected diseases that cause significant burden in areas where treatments do not exist, are inadequate, or are unavailable to the affected populations (e.g., persons in developing countries). Develop strategies to increase access to and improve effectiveness of currently available control interventions for neglected diseases. Develop strategies to increase use and application of social demographic indicators to ensure assessment of health disparities in subpopulations. Develop surveillance systems for zoonoses that incorporate environmental surveillance (e.g., prevalence in vectors and/or vertebrate hosts) to detect pathogens and estimate risk. Evaluate and improve vector management methods, including development of novel pesticides and delivery systems. Develop and enhance surveillance systems for migrating populations, immigrants, refugees, and travelers.

3. Pharmacoepidemiology of Infectious and Other Disease Therapy

Use pharmacoepidemiology (i.e., the study of the utilization and effects of drugs and related medical devices in large numbers of people based on principles of pharmacology and epidemiology) to inform the rational use of pharmaceuticals (e.g., drugs, biologicals, and drug-related devices) for the prevention, treatment, and control of infectious and other diseases [9, 10].

Examples of Priority Research: Determine patterns of pharmaceutical use to evaluate their effect on disease prevention and control and to measure their economic impact. Use pharmacoepidemiologic data to evaluate the effectiveness and safety of disease prevention and control interventions. Enhance pre-marketing study data on safety and effectiveness of pharmaceuticals by conducting appropriate post-marketing surveillance and pharmacoepidemiologic studies to quantify the incidence of known adverse and beneficial effects of pharmaceutical therapy in diverse populations; these efforts should include populations that typically are not assessed in pre-marketing studies (e.g., the elderly, children, and pregnant women). Determine whether the effects of pharmaceuticals are modified by other drugs, unrelated illnesses (i.e., illnesses that are not the target of the therapy), or human genetic factors, and determine the magnitude of the effect relative to alternative interventions used for the same indication. Identify previously undetected adverse and beneficial effects of pharmaceuticals that may be uncommon or delayed.

4. Infectious Disease Elimination

Develop methods to eliminate or eradicate identified priority infectious diseases.

Examples of Priority Research: Assess disease burdens, and define and evaluate strategies and implementation targets for diseases with potential for elimination (e.g., polio, measles, syphilis, tuberculosis and those that are vaccine preventable). Develop new tools for diagnosis, treatment, program monitoring, and surveillance to achieve elimination goals. Evaluate novel approaches to improve the implementation of program interventions and to enhance population coverage for those interventions. Conduct demonstration projects to determine the feasibility of disease elimination, and analyze the public health impact of disease elimination efforts. Evaluate and develop social mobilization and communication methods. Define opportunities to sustain disease elimination efforts through collaboration and integration with other disease-control programs and activities. Investigate the economic impact of priority disease elimination, and estimate the cost-effectiveness of selected prevention and control strategies. Develop optimal strategies for surveillance, and validate criteria for certification of the elimination of priority diseases.

D. VACCINES AND IMMUNIZATION PROGRAMS

Vaccines have been hailed as one of the 10 most important public health contributions of the 20th century [11, 12]. Vaccines currently are used to protect people of all ages from many infectious diseases (e.g., measles, polio, and influenza), and new vaccines are being developed to protect against acute and chronic diseases (e.g., cervical cancer and acquired immunodeficiency syndrome [AIDS]) and against diseases with pandemic potential. To achieve maximum infectious disease protection for all children, adolescents, and adults living in the United States, the following vaccine-associated research efforts must be undertaken: a) the identification and development of suitable vaccine-delivery mechanisms that will reduce health disparities by ensuring the provision of vaccine to all U.S. residents, regardless of socioeconomic status; b) the development of methods that contribute to vaccine safety; c) the conduct of epidemiologic studies to identify populations most in need of vaccine (based on disease burden, risk, and vaccination coverage in a population); d) the assessment of the relative contribution of different modes of disease transmission; e) the determination of vaccination effectiveness in adults and children; f) the development of strategies to ensure full supply and distribution of vaccines to all persons wanting to be vaccinated; and g) the evaluation of the impact of programs created to prevent hospitalization and death in vulnerable populations (e.g., young children and the elderly).

> Our nation can have the highest level of disease protection by ensuring that all Americans have access to a safe and abundant vaccine supply.

1. Immunization Services Delivery Research

Develop, evaluate, implement, and disseminate effective strategies to encourage all persons living in the United States to seek recommended vaccination, and optimize vaccine delivery.

Examples of Priority Research: Develop, evaluate, and monitor effective vaccination programs, especially for new vaccines and those involving new populations of people to be vaccinated (e.g., adolescents and persons in high-risk groups). Develop programs that reduce health disparities in vaccine coverage. Research the efficacy and effectiveness of evidence-based strategies as outlined in the Guide to Community Preventive Services [13].

Conduct health economics research of vaccine programs. Identify strategies to match vaccine supply and demand. Determine the potential usefulness of tailored interventions. Assess the accuracy and usefulness of vaccine registries. Examine the effect of vaccine- associated communication with various audiences (e.g., the public, policymakers, healthcare providers, and health-care organizations) on immunization coverage rates. Test the potential usefulness of alternative settings for vaccine administration (e.g., emergency departments)

2. Vaccine Epidemiology and Surveillance

Enhance epidemiologic research and improve surveillance to better define the burden of vaccine-preventable diseases (VPDs) and to develop more efficient coverage strategies for both new and existing vaccines.

Examples of Priority Research: Determine the burden of disease in specific populations and at-risk groups, and develop new surveillance methods to accomplish this task. Determine whether diseases in specific populations are best prevented by selectively vaccinating those at greatest risk, or indirect infection-control strategies (e.g., vaccinating people not considered to be at high risk). Determine the optimal timing of revaccination for persons with waning immunity (e.g., adolescents and the elderly). Determine the potential risks and benefits of universal vaccination recommendations for specific vaccines. Develop and evaluate novel methods of monitoring and controlling VPDs. Determine the relative effectiveness of mass vaccination against emerging diseases. Determine the risks and benefits of live attenuated virus vaccines. Develop rapid, inexpensive testing for serologic types of VPD strains to measure potential for diminished effectiveness of vaccine control.

3. Vaccine Safety

Develop, implement, and evaluate methods for ensuring safe vaccines and vaccine administration and effective communication of the benefits and risks of vaccines.

Examples of Priority Research: Develop, implement, and evaluate methods to improve detection and assess causality of adverse events following vaccination (e.g., through improved vaccine-safety surveillance and epidemiologic studies). Inform immunization-related decisions by conducting clinical- and population-based research on the pathophysiology and risk factors that influence vaccine-related adverse effects (e.g., individual genetic variation). Develop, implement, and evaluate international standards in vaccine-safety research for use in surveillance systems, clinical trials, and epidemiologic studies to enable data comparability. Evaluate current health-care provider practices regarding prevention, identification, and reporting of vaccine- associated adverse events. Determine behavioral, social, and other factors that impact vaccination acceptance rates among subpopulations, and develop evidence-based methods of addressing potential concerns. Identify and evaluate methods of effectively communicating vaccine benefits and risks to different audiences (e.g., the public and health-care providers). Develop and test promising technologies to enhance vaccine administration safety; evaluate the cost-effectiveness of these technologies, and implement those methods and partnerships deemed to be most cost-effective to monitor, evaluate, and improve vaccine safety.

4. Vaccine Supply

Develop strategies to ensure sufficient supply and appropriate distribution of vaccines.

Examples of Priority Research: Develop and evaluate strategies for raising awareness of the need for and benefits of vaccination, particularly for adults, adolescents, and underserved racial/ethnic groups. Develop a prioritization system for new vaccines based on disease burden, economic analysis of existing markets, and new product availability. Develop and evaluate regulatory strategies for streamlining the review and approval of new vaccines without compromising vaccine safety or efficacy. Collaborate with the National Institutes of Health to develop more efficient means of vaccine production. Develop, evaluate, and implement strategies to more efficiently track vaccine supply from manufacturer to provider. Devise and test strategies to encourage industry interest in manufacturing vaccines. Devise and test strategies to maintain multiple vaccine suppliers to ensure sufficient vaccine inventory. Evaluate population effectiveness of antigen- sparing strategies.

E. BEHAVIORAL, SOCIAL, AND ECONOMIC RESEARCH IN INFECTIOUS DISEASES

Infectious diseases can significantly impact quality of life, society, and all levels of the economy (i.e., local, regional, national, and global). Human behavior plays a critical role in the emergence, spread, and control of almost all infectious diseases [2,14]. However, to reduce the infectious disease burden in the United States, research is needed to further understand the interplay between human behavior, health-associated economics, and the control of infectious diseases. Behavioral interventions to promote adherence to prevention are needed. Because the incidence of infectious disease varies among populations and sub-populations, additional population-specific research focusing on the behavioral, social, and economic factors that contribute to U.S. health disparities is needed. Such research could help identify specific strategies for effective interventions, particularly in populations that are disproportionately impacted by infectious diseases. The conduct of these types of research will help elucidate the causes of health disparities and help guide policymakers in effectively using limited health resources to narrow gaps in health status among populations. New types of research have been identified that show promise for a) identifying effective interventions to prevent the emergence and spread of infectious diseases; b) improving life expectancy and quality of life; and c) improving the economies of societies, communities, groups, and individuals. This research can accelerate health improvements for persons at high risk for illness and help ensure that the greatest numbers of people are protected from infectious threats.

1. Behavioral and Prevention Research to Promote Health

Develop, evaluate, and implement existing and emerging infectious disease-specific behavioral and social science interventions, public health education programs, and health communication research.

Examples of Priority Research: Identify and characterize behaviors and specific determinants (e.g., international travel, sexual behavior, health-related stigma, poverty, and access to healthcare) that put people at risk for routine infectious diseases and those associated with large outbreaks. Examine the physical, social, and cultural factors at the individual, community, and institutional level that directly or indirectly influence a person's risk for infection and for transmitting disease. Engage in community-based participatory research (i.e., a collaborative research process between researchers and community representatives) to improve the effectiveness of behavioral, communication, and health interventions, which can help reduce infection-associated risk. Identify, understand, and characterize health-care-provider behavior and institutional policy associated with a) the containment, spread, and treatment of disease in health-care settings; and b) the implementation of effective strategies to prevent disease transmission and improve health outcomes. Study the effectiveness and practice of culturally competent services and infection-control measures (e.g., hand washing and safe needle use and disposal). Evaluate the effectiveness and sustainability of a) behavioral, health promotion, communication, and prevention programs; and b) interventions for specific populations at various levels (e.g., individual, community, and institutional). Develop evaluation measurement tools to determine the effectiveness of behavioral, health promotion, communication and prevention programs. Evaluate methods for the dissemination and application of research outcomes, model programs, and interventions that have been demonstrated to work. Measure the relative effectiveness of health education programs focused on family planning and preconception health (e.g., those that encourage condom and microbicide use, abstinence, and vaccination against sexually transmitted diseases). Develop new tools and strategies for prevention.

2. Economic Analyses of Infectious Diseases

Measure the economic burden caused by infectious diseases and the economic costs and benefits associated with new and existing prevention interventions.

Examples of Priority Research: Improve understanding of the economics of controlling and preventing infectious diseases through enhanced information and analysis of the economic burden imposed by these diseases. Measure the costs of delivering existing prevention interventions (e.g., vaccines). Conduct cost-effectiveness analyses of existing and novel interventions to allow for direct comparison of interventions. Integrate existing mortality and morbidity estimates and other data to estimate the economic burden caused by all infectious diseases each year in the United States, including travel and health-care- associated infections.

Each year, sample the population to determine estimates of the number of persons who become infected with specific diseases (e.g., influenza) and the burden imposed by these illnesses; these data will improve efforts to target disease-control interventions (particularly during supply shortages and naturally occurring catastrophes).

Measure the cost efficiency of vaccinating or treating different segments of the population (e.g., children and the elderly) in different settings (e.g., physician offices and community mass immunization locations). Apply standardized methodology to evaluate the cost- effectiveness of existing and proposed public health interventions to allow policymakers to make direct comparisons of intervention options. Use econometric analysis to assess the demand for and the supply of infectious disease interventions.

Our genes determine not only the color of our eyes but how our bodies will respond to infection. Learning more about the relationship among genes, the immune system, and infection can lead to better treatments, faster recovery, and longer healthier lives for all Americans.

F. HOST-AGENT INTERACTIONS

Infectious diseases are among the leading causes of death worldwide [15]; in 1998, deaths from infectious diseases were estimated at 13.3 million, or 25% of all deaths. This number likely is an underestimate, however, because deaths from cancer, cardiovascular disease, and respiratory/digestive illness can also be caused by infections (16). Understanding host immune, genetic, genomic, and other factors (e.g., cellular events and immune response mechanisms) is critical to the prevention of illness, disability, and death caused by infectious disease. The identification of genetic and immunologic characteristics that predispose persons to adverse vaccine or drug reactions and the knowledge that numerous chronic diseases are actually caused by preventable or treatable infections also can further efforts to develop new and improved prevention and control modalities. Genetics, race/ethnicity, culture, socioeconomics, sex, education, and behavior influence the outcomes of infection and its associated chronic illnesses. Persons in racial and ethnic minority groups and those who are disadvantaged or undereducated are often most vulnerable to these diseases, which results in substantial health disparities.

Infection that occurs early in life can determine the later chronic health outcomes of individuals and their offspring (17). Additional research in the field of host-agent interactions could eventually enable persons to enjoy additional years of healthy life by delaying death and the onset of illness and disability caused by infectious diseases and any associated chronic conditions.

1. Applied Genomics in Infectious Diseases

Investigate genetic and immunologic features of human-pathogen interactions to understand and prevent infectious causes of acute and chronic diseases.

Examples of Priority Research: Investigate the relationship among human and pathogen genetic and immunologic factors in disease susceptibility, transmission, and natural history; amenability to prophylaxis; response to treatment; priority groups for intervention (i.e., those at high risk); side effects of drugs and vaccines; adverse long-term sequelae; and drug resistance.

2. Infectious Disease and Chronic Disease Associations

Develop improved methods to determine, measure, monitor, treat, and prevent chronic diseases attributable to infectious agents in broad and specific populations.

Examples of Priority Research: Investigate, define, and quantify chronic diseases attributable to infectious agents, particularly those with modifiable, confirmed, and still unproven etiologies. Define the epidemiology of infections and chronic diseases in the presence and absence of infection and other cofactors. Enhance laboratory capacity to detect and monitor biomarkers of infection and of progression to infection-related chronic diseases. Identify populations at risk for infection and chronic health outcomes, and identify factors that increase risk (e.g., socioeconomics, culture, genetics, sexual behavior, non-infectious environmental exposures, and micronutrient status). Develop appropriate interventions that will be readily accepted and used by at-risk populations. Monitor and demonstrate the evolving impact of intervention strategies on infections and their chronic outcomes, including chronic illness and disability. Monitor the indirect effects of interventions on other human-microbe interactions (e.g., antimicrobial resistance, reduced beneficial effects of microbes, and imbalances between microbes and probio tics). Determine when and how modifiable infections influence chronic conditions characterized by inflammation (e.g., cardiovascular and respiratory diseases). Develop strategies for investigating poorly-defined and under-researched syndromes.

G. SPECIAL POPULATIONS AND INFECTIOUS DISEASES

Some U.S. populations (e.g., men who have sex with men, persons who are elderly, and those who are homeless) are at increased risk for being adversely affected by some infectious diseases, which leads to health disparities [2,14]. These vulnerable populations are heavily dependent on public health programs and services for their well-being. Although each vulnerable su-population can be small in size, as a group these sub- populations represent a substantial number of persons who need assistance from public health programs. Pregnant women comprise one such vulnerable group. Infection is a leading cause of hospitalization during pregnancy, pre-term delivery, and postpartum complications [18]. Maternal infection can lead to life-threatening bloodstream infections, prematurity, low birth weight, long-term disability, chronic respiratory and liver disease, and neonatal death [19]. Certain settings have high concentrations of at- risk populations, including nursing homes, correctional facilities, and homeless shelters [20]. Infectious diseases that emerge from such settings or within these populations can eventually spread to the general population. Fortunately, solutions to some of the health problems that affect one group often can be used to manage different diseases in other at-risk populations. Research focusing on ameliorating adverse health effects experienced by vulnerable populations would better the health of not only those who are most vulnerable, but persons in other populations, including those that are not considered to be high-risk. Translating research into practice among high-risk populations is particularly needed. In the long run, this research would enable all persons, regardless of risk level, to enjoy additional years of healthy life by delaying death and the onset of illness and disability.

> We can improve the health of our nation by ensuring that our most vulnerable populations—victims of disasters, pregnant women, the very young, and the very old—are well protected against infectious diseases.

1. Health Disparities and Infectious Diseases

Develop, evaluate, and implement strategies to reduce and eliminate health disparities associated with infectious diseases.

Examples of Priority Research: Develop, evaluate, and implement strategies to reduce and eliminate disparities associated with infectious diseases that affect certain groups of persons (e.g., those of a certain age, sex, and race/ethnicity; those with a lower education and income level; those who engage in certain cultural practices; and those who have certain religious beliefs, disabilities, sexual behaviors, geographic locations, occupations, and languages). Investigate the protective and risk factors for infectious diseases for these and other subgroups of the larger population (e.g., popular cultural customs). Investigate the relative efficacy of new diagnostic and screenin approaches to detect infectious diseases across populations. Develop and evaluate new tools and materials for reducing rates of infectious disease based on segmented cultural approaches, education, communication, and prevention.

2. Infectious Diseases Among Populations at High Risk

Develop, evaluate, and implement strategies to assess the risks associated with infectious diseases, and better monitor infectious disease rates to reduce the impact of these diseases in populations at high risk.

Examples of Priority Research: Identify and evaluate prevention and control interventions in the following four situations: a) adverse environments, b) highly susceptible hosts, c) special risk-factor groups, and d) high-density populations. Specific environments that should be further studied include war and disaster zones, detention and correctional facilities, and rural and low population frontier areas. Particularly high risk groups include recent immigrants, refugees, international travelers, the homeless and incarcerated, regular or occasional drug and alcohol users (particularly persons who use methamphetamines), men who have sex with men, persons with immunodeficiencies, and persons who have undergone solid-organ transplants. Evaluate changes in infectious disease risk across the lifespan, with emphasis on the very young and elderly populations.

3. Perinatal and Neonatal Infectious Diseases

Develop point-of-care diagnostic methods, characterize strategies for preventing perinatal infections, and assess the health impact of existing prevention tools.

Examples of Priority Research: Identify, characterize, and assess effective strategies to prevent and control perinatal infections. These strategies will help ensure healthy pregnancies for both mothers and infants and prevent premature births, birth defects, and other morbidity and mortality that can affect the fetus, infant, and older child. Examine the relationship between particular maternal and infant infections, and determine the long-term outcomes for infected infants. Through interdisciplinary research, examine infectious processes, inflammatory responses, reproductive health, and genomics to identify novel strategies for preventing premature births. Develop point-ofcare diagnostic methods for use in prenatal and obstetric health-care settings for various infections. Identify and evaluate effective primary prevention strategies through health services research (e.g., maternal and infant screening, education, and treatment).

Chapter IV

PROMOTE PREPAREDNESS
TO PROTECT HEALTH

September 11, 2001, the day thousands of civilians fell victim to an extreme act of terror on U.S. soil, marked a tragic day in the lives of Americans and served as a wake-up call for public health professionals nationwide. In addition to acts of terrorism, the United States recently has been severely impacted by natural disasters. In September 2005, two Hurricanes—Katrina and Rita—pounded America's Gulf Coast killing hundreds of people, destroying homes and property, and decimating the health-care and public health infrastructure [21].

Public health emergencies, including acts of terrorism, natural disasters, and infectious disease outbreaks, pose challenges for all communities. Public health organizations at all levels and in all nations must be adequately prepared for these threats. In 2003, when six state and local health departments and three provincial governments were asked how prepared their respective states were to respond to bioterrorism or other emergency threats, they responded that although the states are "off to a good start," more work must be done [22]. Public health agencies must develop infrastructures that improve their capacity to respond to any type of emergency [23]. This all-hazards approach will help ensure that all sectors of society are prepared for any public health emergency [24]. Public health systems must be able to anticipate, recognize, and respond to large- scale disturbances, whether they are intentional acts of terrorism or naturally occurring events. Although emergency public health has always been a public health activity, health services research in preparedness and response must be made a priority; focus must be given to preparation for emergency events, including planning, tracking, and response [23]. Adequate preparation for public health emergencies depends on state coordination in cooperation with community, local, tribal, and federal officials.

Research on community preparedness and response is needed to minimize the impact of public health emergencies on affected communities and to promote health. Research on detection and diagnosis, health communication, and public-health workforce training will contribute to existing public health efforts to maintain safe and healthy communities in the face of a wide range of public health emergencies. Such research will help achieve the goals of protecting the health of individuals and communities in the 21st century, particularly those that are often disproportionately affected by injury and illness, including ethnic and cultural minorities, persons with disabilities, and those in low socioeconomic strata [23]. Research categories in this chapter include:

A. Vulnerable Communities and Populations
B. Infrastructure and Prevention
C. Public Health Workforce Preparation and Front-line Prevention and Response
D. Detection and Diagnosis of the Hazards and Medical Consequences Associated with Emergency Events
E. Communications
F. Community Preparedness and Response Improvement

A. VULNERABLE COMMUNITIES AND POPULATIONS

Public health emergencies (e.g., catastrophic events such as acts of terrorism, natural disasters, and infectious disease outbreaks) pose challenges for all communities. However, disaster impact is distributed unevenly in the population based on social, health, psychosocial, or economic status. Vulnerable communities must be able to maximize their preparedness to ensure recovery from and resilience to public health emergencies. It is the role of government at all levels to prevent and mitigate where the impact is greatest. Reducing or mitigating threats among diverse populations is challenging, because the responses, interpretations, judgments, and self-protective behaviors of persons in vulnerable groups are influenced by different cultural, social, and environmental forces; the characteristics associated with the public health emergencies themselves (e.g., breakdown of social order) also affect these behaviors. The risk and protective factors of these vulnerable populations should be the focus of research and programmatic efforts. Standardized and reproducible measurements and methods must be developed to describe and quantify the vulnerability of communities. Advances in knowledge regarding which demographic, behavioral, and health risk factors make communities more vulnerable will enable the identification of strategies and interventions to reduce communities' susceptibilities across a broad range of hazards. Development and application of comprehensive models to prioritize preparedness and response activities at the community level are particularly critical [25].

Research focusing on the populations most likely to be negatively impacted by catastrophic events will further existing public health efforts to keep communities protected from environmental, occupational, infectious, and terrorist threats. This research should be conducted in collaboration with community stakeholders and academic, private, tribal, and government researchers.

> Communities disconnected by poverty, language, and geography suffer most when disaster strikes. Identifying those at greatest risk will ensure that all Americans are better prepared and protected.

1. Determinants of Community Vulnerability to Public Health Emergencies

Identify, measure, and understand the factors and dynamic processes that influence the vulnerability of defined community populations.

Examples of Priority Research: Develop reliable, valid tools and strategies to profile the vulnerability of communities along multiple sociocultural and community dimensions, including the mechanisms responsible for health disparities. Explore approaches to effectively intervene in the mechanisms or pathways that link sociocultural, physical, psychological, pre-existing health, and economic characteristics to community vulnerabilities during public health emergencies to reduce negative outcomes. Evaluate the cumulative effect of exposure to multiple adverse events over time on the vulnerability of communities and the variability in their ability to adaptively respond and recover. Develop rapid assessment tools and more detailed and enduring evaluation tools to measure community vulnerability.

2. Risk Appraisal and Adaptive Behavior During a Public Health Emergency

Explore how local influences, cultural factors, and past experiences affect the perception of risk among individuals and their communities and shape their behavioral responses.

Examples of Priority Research: Identify the influence of social, cultural, and historical factors on perceptions of risk and adaptive responses associated with public health emergencies, especially among vulnerable, high-risk populations. Determine possible sociocultural barriers to and facilitators of healthy adaptive response patterns among individuals and communities. Evaluate the way perception and appraisal of risks influence individual and group interpretations of that risk and their subsequent behaviors, and determine the effectiveness of communications from public health professionals concerning ways to reduce these risks. Determine the influence of ethnicity and culture in a community's request for help and the availability of help. Evaluate and determine the most effective means of developing effective intervention strategies in these communities, including technical assistance to tribal and minority communities.

3. Predictive Strategies for Risk and Recovery in Vulnerable Populations after a Public Health Emergency

Identify, evaluate, and model the components of risk, including hazards, vulnerabilities, and resources.

Examples of Priority Research: Understand how ongoing and multiple risk or protective factors and dynamic processes interact to affect vulnerability and/or resilience before, during, and after public health emergencies. Identify models to incorporate pre-existing dynamics, exposures, and other conditions within vulnerable communities that influence the probability and severity of negative outcomes during an event.

4. Assessment Strategies for Populations Affected by Public Health Emergencies

Identify, develop, and evaluate strategies and policies to prevent, mitigate, and treat adverse health effects.

Examples of Priority Research: Evaluate methods of assisting every segment within a community in establishing unique plans and policies that will assess and prioritize each segment's needs and outline available resources to be relied upon in times of crisis.

5. Public Health Emergency Response Strategies

Develop strategies to adapt the public health response system to accommodate heterogeneous social and physical contexts.

Examples of Priority Research: Develop, examine, and evaluate procedures and systems for identifying the elements that distinguish one subgroup from another; these activities will facilitate the development of community-based systems for provision of services to each of these groups. Develop methods to comprehensively assess the needs of each population group, understand how best to satisfy those needs, and modify response strategies accordingly.

6. Public Health Emergency Management Strategies

Identify, develop, and assess comprehensive risk-management strategies for heterogeneous populations.

Examples of Priority Research: Identify potential causes of adverse physical and psychological health effects and the segments of the population likely to be affected by natural or man-made disaster events. Determine the means for categorizing stressors (both real and perceived) and rectifying the effects of each stressor for specific populations. Priorto a public health emergency, study ways to identify and select risk- control activities and methods of assigning responsibility to ensure that all aspects of the response are directed appropriately and effectively.

B. INFRASTRUCTURE AND PREVENTION

U.S. public health systems must be prepared to anticipate, recognize, and respond to large-scale natural catastrophes, intentional acts of terrorism, and unintentional manmade disasters. In national forums, Congressional testimonies, and in their communities, local public health officials have stated that strengthening overall local public health practice is the best way to address the public health consequences of bioterrorism [22]. The National Association for County and City Health Officials has encouraged the use of community-wide surveillance systems capable of detecting bioterrorist events and other health emergencies [22].

Efforts to strengthen the public health infrastructure will enable the U.S. public health system to be more resilient and better prepared to detect and respond quickly to potential or actual health emergencies. Specifically, operational and applied research is needed to promote and evaluate integrated systems of care and risk management, incident management, and communication among health and safety authorities and residents. Research also is needed to identify the contextual, incident management, and infrastructural variables that lead to model integration practices, community and border response systems, and best practices for health-related government and public information exchange.

Finally, research must include assessment of populations who may be disproportionately impacted by emergency events because of certain factors (e.g., limited access to care, geographic barriers, poor health status, and socioeconomic disadvantages).

> Our nation's emergency response system must have timely, accurate information to use in the development and application of intervention strategies to help public health systems recover.

1. Critical Infrastructure Systems and Processes

Assess and prioritize methods and practices to better protect public health systems that are critical for maintaining healthy populations.

Examples of Priority Research: Identify the critical public health systems that are frequently disrupted during public health emergencies (e.g., food and water supply and their safety, waste management, vector control, and veterinary services). Create or strengthen guidelines, practices, and collaborations that will ensure that health-sustaining public health systems and preventive practices are functional before, during and after public health emergencies.

2. Public Health and Clinical Response Systems

Assess and identify strategies and model practices for the integration of public health and clinical response systems during public health emergencies.

Examples of Priority Research: Assess health-care utilization and surge capacity (e.g., the availability of beds, staff, operations, logistics, decontamination, antiviral drugs, ventilators, diagnostic imaging technology, hand-hygiene products, personal protective equipment intended to be used by the general population, operating rooms, and medical supplies) for different scenarios (e.g., conventional weapon terrorism, hurricanes, tornadoes, earthquakes, tsunamis, airplane or railway crashes, seasonal and pandemic influenza outbreaks, and vaccine shortages). Develop and test screening and triage strategies to a) assist resource allocation efforts; b) inform medical triage decisions; and c) ensure comprehensive care and follow-up associated with mass casualty events at the local level. Determine whether systems integration can a) reduce the impact of surge capacity on already fragile, overburdened systems of trauma and emergency care and b) influence the efficacy of infection control interventions (e.g., respiratory hygiene and universal masks) in limiting dissemination of respiratory infections in various settings (e.g., hospitals, emergency departments, outpatient clinics, and long-term care facilities). Evaluate recovery and maintenance capacity for non-event-related health care during emergencies and recovery of core public health functions after a public health emergency. Assess the efficiency and effectiveness of syndromic surveillance, field triage, and other early detection systems in a) improving survival and outcomes for persons sustaining trauma and b) identifying and isolating patients with pandemic influenza or other emerging infectious diseases in health-care settings.

Evaluate community-based mental-health care aimed to assist people in coping with the consequences of various public health emergencies. Identify and evaluate alternative partners (e.g., community health centers, tribal health centers, workplace health systems, and school-based health services) that can be relied on to help coordinate public health functions and reach minority populations and communities.

3. Human Migration, Mobility, and Quarantine Issues Associated with Public Health Emergencies

Assess and evaluate the role of human migration in the public health response to public health emergencies occurring around the world.

Examples of Priority Research: Identify risk and protective factors and evaluate interventions designed to prevent introduction of, and control the spread of disease, social deterioration, injury, violence, disability, and death associated with human migration and global mobility. Carry out operational research to improve standards, guidelines, and practices to prevent disease, injury, and disability among displaced populations in temporary shelters, quarantine facilities, long-term evacuation centers, and new permanent living spaces. Collaborate with governments, agencies, and partner organizations to improve the national and international communities' understanding of and response to complex and humanitarian emergencies; such research should also focus on a) the diverse health issues experienced by refugees, evacuees, and other displaced populations; b) the medical screening/detection strategies and health protection interventions in place at international borders, ports of entry, and local communities impacted by an influx of new populations; and c) the resiliency of the medical care and public health service delivery infrastructure when faced with displaced populations or communities whose normal movement is restricted as a result of epidemic disease, a natural or technological disaster, or armed conflict.

4. Community Actions in Public Health Emergencies

Describe and explain how diverse communities detect, interpret, respond to, and communicate perceived and actual public health threats.

Examples of Priority Research: Examine critical social processes and reporting relationships that are relied on by communities and population groups to enhance hazard awareness and trigger prevention or mitigation efforts. Evaluate the health risks and unique needs of vulnerable populations, including immigrants and those with limited English proficiency, during and after adverse events. Evaluate community and faith- based organizational capacity to respond to and provide emergency support during crisis events. Assess the impact of community-based social distancing measures (e.g. school closures, cancellation of public events, and altered use of public transportation). Evaluate the willingness and ability of the public to adopt personal protective measures (e.g., hand hygiene, cough etiquette, and decreased face-hand contact). Assess public perception of risk versus benefit concerning the use of these measures and public understanding of and access to resources, particularly for populations impacted by health disparities.

5. Local and Regional Operational Strategies for Managing Public Health Emergencies

Optimize local and regional operational strategies for information exchange, decision- making, and command and control, and define the optimal roles and functions for persons involved in managing public health emergencies.

Examples of Priority Research: To ensure public health and safety, evaluate systems currently used to promote effective information exchange and to establish lines of authority, responsibility, and accountability across multiple agencies and jurisdictions.

C. PUBLIC HEALTH WORKFORCE PREPARATION AND FRONT-LINE PREVENTION AND RESPONSE

The public health workforce, comprised of approximately 450,000 professionals and other front-line responders [26], must be adequately prepared for prevention and response in the event of a large-scale public health emergency (i.e., catastrophic events such as acts of terrorism, natural disasters, and infectious disease outbreaks). Public health professionals (which include not only persons employed in a public health or health-care setting, but other professionals who are or can become involved in public health activities) must be able to rapidly assess the health of populations in affected areas; implement surveillance systems to monitor the health status of populations following an event; produce and distribute health education materials to medical providers, affected populations, and others; control disease outbreaks and identify and mitigate any resulting environmental health hazards; and provide for the needs of special populations (e.g., the disabled and elderly) during and after a disaster event [27]. All front-line public health practitioners should have the skills and abilities to recognize a time of crisis or emergency and to intervene by helping to prevent disease, disability, and injury. Research is needed to understand the importance and uses of cultural competency training and the recruitment of professionals representative of the population. Determining the proficiencies required during emergency response and recovery may require identifying specific skill sets, competencies, or cross-training to ensure that public health professionals can adapt to rapidly changing community health needs during the evolution of an emergency. Research focusing on public health preparedness and response should be conducted in collaboration with community stakeholders and academic, private, and government researchers.

1. Community and Regional Response During Public Health Emergencies

Ensure scientific rigor in the design, implementation, and evaluation of drills and exercises.

Examples of Priority Research: Assess and evaluate strategies for modeling community- and regional-level response to test system capacity and capability to rapidly and effectively recover essential infrastructure functions (e.g., communications and power supply).

2. Support for Front-line Personnel Involved in Health Protection Functions During Public Health Emergencies

Improve and evaluate the countermeasures, personal protective equipment, and health policy guidelines that support members of the public health workforce and maximize worker safety, personal resilience, self-confidence, and performance.

Examples of Priority Research: Develop appropriate medical screening and monitoring programs for workers involved in rescue and recovery at scenes of chemical, biological, radiological, and nuclear threats. Gather information and recommendations on the performance, availability, and appropriateness of personal protective equipment. Develop recommendations for management processes and worker training to prevent injury and illness among first responders and recovery workers at the local, state, and federal levels. Assess technologies designed to better protect emergency responders and other relief or recovery workers. Develop guidelines for emergency responders regarding the selection, use, and maintenance of respirators and other personal protective equipment in disaster situations.

3. Proficiency of the Public Health Workforce in the Event of Disaster

Define and assess the knowledge, attitudes, and proficiencies needed by the public health workforce to successfully ensure the safety and well-being of the community.

Examples of Priority Research: Develop ways to ensure the timely and continual updating of the standards and competencies necessary for an effective, efficient health workforce. Develop methods for ongoing assessment of the training and development needs of the public health workforce. Evaluate strategies for recruitment and retention in maintaining the workforce, particularly health workers belonging to minority groups.

D. DETECTION AND DIAGNOSIS OF THE HAZARDS AND MEDICAL CONSEQUENCES ASSOCIATED WITH EMERGENCY EVENTS

Detection and diagnosis are essential components of the public health and medical response needed to mitigate the impact of a public health emergency (i.e., catastrophic events such as acts of terrorism, natural disasters, and infectious disease outbreaks) on the community. Research that focuses on enhancing the ability of responders to rapidly detect and accurately diagnose the etiology of adverse health events is needed. Specific research areas include a) the evaluation and integration of public health and medical surveillance and reporting systems; b) the exploration and leveraging of rapid identification advancements in the fields of medical diagnostics and environmental detection; c) the development and evaluation of strategies and methods for immediate impact assessment (e.g., exposure modeling and environmental contamination and decontamination assessment methods) and long-term community impact evaluation (e.g., health registries for long-term health status follow-up and full community impact assessment); and d) evaluation of strategies for consequence management. Systems must be evaluated for their flexibility in collecting and compiling health data of all types and for reaching communities that may disproportionately experience the affects of public health emergencies. This research should be conducted in collaboration with community stakeholders and academic, private, and government researchers.

> Advanced information systems provide real-time information to support real-time responses to public health threats

1. Health Surveillance Systems Involved in Public Health Emergencies

Develop and integrate systems to detect, report, and investigate illness and injury associated with intentional and unintentional emergent health threats.

Examples of Priority Research: Identify and assess potential health-indicator data sources. Validate, standardize, and integrate public health and medical surveillance systems to improve capabilities for the detection of natural and man-made threats to public health.

2. Rapid Clinical Diagnostic Capabilities During Public Health Emergencies

Develop or enhance rapid clinical diagnostic capabilities to identify significant exposures to potentially hazardous agents.

Examples of Priority Research: Explore and evaluate modern techniques for and advancements in the rapid clinical diagnosis and identification of significant etiologic agents, and examine ways to leverage these diagnostic and exposure assessment capabilities within public health and medical laboratory systems.

3. Environmental Detection and Decontamination During Public Health Emergencies

Develop or enhance methods of rapidly detecting, identifying, and decontaminating persons and environments that may be adversely impacted by an emergency event and could potentially pose a threat to others.

Examples of Priority Research: Explore and evaluate a) modern techniques and advancements used for rapid identification of environmental health threats and for interpretation of data to assist public health decision-making and accurate health guidance; b) methods for assessing environmental contamination; and c) strategies and methods for evaluating the effectiveness of decontaminating or cleaning up impacted environments or persons to reduce the probability of exposure to health threats.

4. Rapid Assessment of Exposure and Impact Data from Public Health Emergencies

Identify and evaluate model strategies for rapidly assessing exposure impacts and resource needs during adverse health events that impact a community, and evaluate the tools needed to support these strategies for resource allocation and health-status tracking

Examples of Priority Research: Develop practical strategies and increase capacity for rapid collection of appropriate exposure and impact data during (or promptly after) an emergency event to be used for needs assessment, resource allocation, and long-term health status tracking (e.g., health registries). Evaluate data collection practices for ethical approaches immediately following any public health emergency. Assess the longterm benefits of these practices within other scientific and public health disciplines.

E. COMMUNICATIONS

Communication is essential to community preparedness and response associated with any public health emergency (i.e., catastrophic events such as acts of terrorism, natural disasters, and infectious disease outbreaks). Effective communication during a public health emergency requires the delivery of clear health-related messages and close involvement of the affected community [28]. Diverse populations may require tailored approaches to ensure that messages are delivered and accepted appropriately, particularly those that face barriers to traditional modes of communication [28].

Research must focus on communication between the public health system and communities, within the public health system, and between the public health system and emergency responders.

1. Risk Communication and Information Dissemination

Identify and develop effective communication strategies, tools, and mechanisms to facilitate rapid and accurate communications regarding risk information and public health recommendations to affected populations.

Examples of Priority Research: Work with the news media and other organizations and agencies from which people obtain information (e.g., community centers, churches, and schools) to evaluate the effectiveness and impact of various forms of messaging and communication strategies (e.g., content, timing, dissemination, and monitoring of reactions). Review scientific validation, approval, and clearance procedures used by the public health community to ensure the timely communication and dissemination of vital and valid information. Evaluate existing practices to minimize the dissemination of incomplete, incorrect, or potentially harmful misinformation. Ensure that adequate processes are in place to translate health guidance into effective preventive interventions for a diverse population. Investigate the use and effectiveness of culturally competent and language-specific messages in relaying risk-related information and response recommendations.

2. Emergency Response Communications Technology

Optimize strategic communications technology to allow for efficient event response across multiple jurisdictions.

Examples of Priority Research: Develop and test communications technology, and evaluate communication strategies (e.g., use of common language and terminology, interoperability, and redundancy) for their effectiveness in ensuring optimal communication within and among responders, the public health system, and other critical response institutions and organizations during a public health emergency

F. COMMUNITY PREPAREDNESS AND RESPONSE IMPROVEMENT

Outcome measurement ensures that public health agencies remain accountable for their performance. Comparison of public health outcomes allows for the development of best practices and standards for preparedness, response, and recovery. Such measures permit the qualitative and quantitative evaluation of the effectiveness of public health systems during and after emergencies. Responses to outcome assessments can be tracked over time to identify system improvements or changes. Although measuring preparedness against an event that does not regularly occur is inherently difficult, research needs associated with outcome measurement include a) conducting systematic collection and analysis of performance data and b) developing a science base for public health practice, preparedness, response, and recovery improvement.

1. Outcome Measurement for Preparedness Improvement

Develop, evaluate, and apply outcome measures for public health practice, preparedness, response, and recovery improvement.

Examples of Priority Research: Test research methodology for application in a "just-intime" modality to ensure that evaluation of events can be performed prospectively and appropriate data can be collected for future analysis. Evaluate existing preparedness plans for their incorporation of strategies and capabilities intended to help an agency a) prevent, respond to, and recover from acts of terrorism, natural disasters, and other emergencies and b) protect the public. Evaluate public health and health-care delivery service program access and effectiveness (including trauma-care outcomes) during a public health emergency. Identify community knowledge, attitudes, practices, behaviors, and perceptions of needs, desires, and priorities related to health-care practices and public health services designed to mitigate, respond to, or ensure recovery from public health emergencies. Establish and evaluate benchmarks for public health preparedness improvements. Conduct economic analysis of preparedness and response interventions.

Chapter V

PROMOTE HEALTH TO REDUCE CHRONIC DISEASES AND DISABILITY

Many non-infectious conditions, including developmental and other types of disabilities, birth defects, and chronic diseases, continue to impact the health and well-being of Americans throughout all stages of life. Approximately 17% of U.S. children have a developmental disorder, and 54 million Americans of all ages have some type of disability, nearly half of whom face severe and chronic disabling conditions. In addition, one in every 33 babies in the United States is born with one or more disabling birth defects, which can take a lifelong toll on the emotional, medical, financial, and social well-being of patients and their families. Chronic conditions remain the leading cause of illness, disability, and death in the United States: 31% of children have a chronic health condition, and nearly 45% of the overall U.S. population has at least one chronic condition, leading to suboptimal quality of life and economic cost.

To improve the health of our nation, efforts to promote the health of persons of all ages living in the United States must focus on the prevention and control of these conditions. The Health Protection Goals, developed by CDC, define life stages by the following age categories: infant and toddlers (ages 0-3 years); children (ages 4-11 years); adolescents (ages 12-19 years); adults (ages 20-49 years); and older adults (ages >50 years). Research is needed to a) promote health and reduce the burden of chronic diseases and comorbidities though effective interventions, including evidence-based early detection and screening; b) develop and disseminate interventions to promote healthy behaviors among women before conception and during pregnancy and among all persons in every stage of life; c) develop and evaluate strategies to ensure optimal child and adolescent development and to identify developmental disabilities early in life; d) reduce the burden and disparity of chronic disease in adults, older adults, and seniors; and e) promote health and prevent illness among persons with disabilities across their lifespan. Such research will foster health and well-being in persons with disabilities, birth defects, and chronic conditions and will help reduce the national prevalence of these often preventable conditions. Research areas in this chapter include:

A. Health Across the Lifespan
B. Infant and Maternal Health
C. Health and Development of Children and Adolescents
D. Adult and Older Adult Health
E. Health Among Persons with Disabilities

A. HEALTH ACROSS THE LIFESPAN

Chronic conditions are the leading cause of illness, disability, and death in the United States [29]. Chronic diseases, including cardiovascular disease, cancer, chronic obstructive pulmonary disease, and diabetes are among the most prevalent, costly, and preventable of all health problems. In the United States, almost 125 million persons (45% of the population) have at least one chronic condition [30], and the medical costs for persons with chronic conditions account for almost 75% of the $1 trillion spent on health care each year in the United States [31]. People with chronic conditions spend six times more per year on health care than those who are not chronically ill, and those who have a chronic condition that results in limitations in day-to-day functioning spend 16 times more than persons who have no chronic illness [30]. Tobacco use and other risk behaviors continue to be the leading contributors to preventable illnesses and causes of death in the United States [32]. In 2000, the leading causes of death were tobacco, poor diet, physical inactivity, and alcohol consumption [32]. The prevalence of persons who are overweight and obese, characteristics that have been associated with increased prevalence of and morbidity from type 2 diabetes, hypertension, arthritis, and some cancers, has more than doubled during the last 40 years [29]. Sustainable strategies to establish appropriate health habits and behaviors during and across all stages of life are needed to promote lifelong mental and physical health and reduce the risk of the leading causes of morbidity and mortality, including chronic diseases (e.g., arthritis, heart disease, stroke, cancer, and type 2 diabetes), injury, violence, tuberculosis, human immunodeficiency virus infections, and other sexually transmitted diseases.

Profound disparities exist among persons in all stages of life; disparity can be seen in health behaviors, risk factors (e.g., tobacco use, oral health, poor diet [including low consumption of fruits and vegetables], and physical inactivity), insurance coverage, access to health care, health outcomes, and disease burden. Understanding the determinants of these disparities and comorbidities is necessary to ensuring effective interventions that address economic, structural, cultural, and individual barriers to optimal health are developed and implemented for diverse population groups and communities.

Research is needed to accelerate the dissemination of effective, sustainable interventions, including evidence-based early detection and screening, to protect and improve health throughout the lifespan. These interventions also can be used to discourage risky behaviors that contribute to excess morbidity and mortality. Research focusing on health maintenance for persons in every stage of life, including palliative care, is needed to promote optimal lifespan and ensure the best possible quality of life. This research will ultimately be used to help support people in all communities in their efforts to achieve healthy and productive lives.

Almost half of all Americans suffer from a chronic illness. Prevention research holds the promise that people, especially those at risk of health disparities, will achieve their optimal lifespan with the best possible quality of health in every stage of life.

---

1. Implement Effective Health Promotion Strategies

Develop and evaluate strategies that enable families, employers, and communities to establish and sustain healthy behaviors across the lifespan.

Examples of Priority Research: Conduct dissemination research to achieve widespread adoption and implementation of proven family-, school-, worksite-, faith-, and community-based interventions to sustain active lifestyles, healthy eating habits (including increased consumption of fruits and vegetables), oral health, tobacco cessation, and mental health. Develop and implement approaches to prevent or delay the onset of chronic diseases and conditions. Conduct research on the prevention of obesity and maintenance of optimal weight throughout the lifespan. Conduct economic research to identify the most promising health promotion interventions. Support high priority research identified through the systematic review process of the Guide to Community Preventive Services [13].

2. Reduce the Burden of Chronic Diseases

Develop and evaluate strategies that enable families, employers, and communities to control and prevent the burden of chronic diseases and to reduce the preventable physical and emotional adverse health effects associated with these diseases and conditions.

Examples of Priority Research: Conduct research to estimate the current and future burden of chronic diseases, and identify risk factors that contribute to the burden, including health behaviors, literacy, education, socioeconomic status, insurance coverage, access to health care, quality of care, self-care and preventive-care practices, and health-systems structure. Develop, implement, and evaluate strategies, including evidence-based early detection and screening, that enable families, employers, and communities to control and prevent chronic diseases (e.g., heart disease, stroke, type 2 diabetes, cancer, and arthritis) and reduce the preventable physical and emotional adverse health effects associated with these diseases and conditions. Examine linkages within the clinical-care system to improve access to and quality of health care. Conduct community-based participatory research to develop and implement strategies that can be used to establish and maintain healthy behaviors throughout the lifespan (i.e., those that protect and sustain health, reduce the risk of chronic disease and disability, and maintain and enhance quality of life); these strategies should be particularly targeted to adults, including those who are elderly.

3. Reduce Disparity in Chronic Disease Burden and Risk Factors

Understand the determinants of disparities in chronic disease burden and risk factors, and implement effective interventions that address economic, structural, cultural, and individual barriers to optimal health, particularly among populations in which health disparities persist.

Examples of Priority Research: Conduct research to estimate the extent and depth of disparities in health behaviors, risk factors (e.g., tobacco use, poor diet, and physical inactivity), insurance coverage, access to health care, health outcomes, and disease burden that exist at all stages of life. Identify and understand the causal factors (e.g., health behaviors, literacy, education, socioeconomic status, insurance coverage, access to health care, quality of care, self-care and preventive-care practices, and health systems structure) that contribute to the significant disparity in chronic disease burden experienced by various populations (e.g., groups characterized by race/ethnicity, gender, sexual orientation, disability, and socioeconomic status). Develop and implement interventions that work across diverse communities and population groups to ensure that all persons, particularly those at high risk for chronic disease, achieve an optimal lifespan and maintain the highest possible quality of life at any age.

B. INFANT AND MATERNAL HEALTH

One in 33 babies born in the United States has one or more severe birth defects [33], and 17% of children <18 years of age have a developmental disability [29], which can have lifelong emotional, medical, financial, and social implications for patients and their families. Approaches to enhance safe motherhood and promote infant and maternal health can significantly improve pregnancy and birth outcomes and help to stop the life-long effects of poor birth outcomes on affected infants and their families.

Profound and persistent health disparities exist in maternal and infant health, pregnancy outcomes, and risk behaviors among women of reproductive age [34]. Eliminating health disparities for mothers and their children is a public health priority; several Healthy People 2010 objectives directly address the need to reduce these health disparities [1]. Research focusing on identifying and implementing promising approaches to reduce these disparities is needed. In addition, effective interventions that promote pregnancy planning, preconception care, and safe motherhood and improve the health of all women and their infants also must be developed, implemented, and evaluated. Such interventions will prevent unintended pregnancy, teen pregnancy, and pre-term delivery; increase access to and quality of care before, during, and after pregnancy; and increase breastfeeding rates. Promoting healthy pregnancy and birth outcomes through the prevention and control of disability across the lifespan is critical to ensuring that infants and toddlers grow into healthy children and adults.

Every child deserves a healthy, strong start in life. Research that supports healthy mothers, healthy pregnancies, and healthy babies can ensure that our smallest citizens have the brightest futures possible.

1. Pregnancy Planning and Preconception Care

Develop and evaluate strategies of promoting pregnancy planning and preconception care to improve birth outcomes and reduce the life-long effects of poor birth outcomes.

Examples of Priority Research: Initiate and assess research-to-practice initiatives to promote pregnancy planning and preconception care through a broad range of healthcare programs and services for women, including women with disabilities. Initiatives should include assessment of maternal and paternal health history and behaviors before, during, and after pregnancy; delivery of vaccinations; screening for disease and genetic disorders; promotion of folic acid use; assessment and treatment of mental health disorders; management of existing illnesses (e.g., diabetes); and provision of health services to address risky behaviors (e.g., smoking, alcohol use, and obesity). Identify indicators of preconception care and women's health that have the greatest impact on birth outcomes.

2. Conditions Associated with Hereditary Birth Defects and Blood Disorders

Identify and evaluate effective interventions to improve the health of persons with hereditary birth defects and blood disorders.

Examples of Priority Research: Assess factors involved in the causation, course, prevalence, and burden of secondary physical, social, emotional, and mental health conditions that affect persons with hereditary birth defects and blood disorders. Identify interventions to maximize the health and well-being of such persons.

3. Healthy Birth Outcomes

Identify and assess the impact of factors that lead to birth defects and developmental disabilities, and identify and evaluate interventions to prevent such adverse events and promote healthy birth outcomes.

Examples of Priority Research: Identify and assess interventions, including general health care for women, that can improve birth outcomes by a) preventing risk of unintended pregnancy, teen pregnancy, and pre-term delivery and b) elucidating the needs of near-term infants (i.e., infants born at 34–3 7 weeks' gestation). Identify and assess the impact of factors that contribute to adverse birth outcomes (e.g., birth defects and developmental disabilities), including those associated with genetics, maternal and paternal environmental exposures, cultural practices, maternal mental health, and socioeconomic status. Identify new risk factors for birth defects and other adverse birth outcomes during pregnancy planning, preconception, pregnancy, and the newborn period. Conduct studies to identify potential interventions to address adverse birth outcomes, including appropriate nutrition and physical activity.

4. Promote Safe Motherhood and Infant Health

Develop and evaluate strategies to promote safe motherhood and promote maternal and infant health.

Examples of Priority Research: Develop and evaluate strategies to enhance infant and maternal health by increasing access to and quality of care before, during, and after pregnancy. The implementation of these strategies will prevent unintended pregnancy, sexually transmitted diseases, teen pregnancy, and pre-term delivery. Understand the determinants of infant feeding behavior, including breastfeeding initiation and continuation and the impact of premature and near-term birth on breastfeeding rates. Conduct clinical trials of drugs to prevent mother-to-child transmission of human immunodeficiency virus (HIV) and other diseases. Identify the role of fetal development and infant care, including increasing breastfeeding practices and improving detection and treatment of maternal mental health disorders, on reducing chronic disease and disabilities in subsequent life stages.

C. HEALTH AND DEVELOPMENT OF CHILDREN AND ADOLESCENTS

Approximately 17% of children in the United States have some type of developmental disorder [29], and an estimated 31% of all children and adolescents have some type of chronic health condition (35,36), placing them at increased risk for poor health and educational outcomes. In addition, over the past 30 years, rates of obesity, a condition that has been associated with early onset of type 2 diabetes and decreased cardiovascular health [37], have increased by two- to three-fold among children and adolescents. Health issues that begin in utero and early in life can have life-long consequences. Therefore, the development and dissemination of effective interventions that ensure early access to appropriate services likely would significantly reduce excess morbidity and mortality among children, adolescents, and adults and would result in improved health and educational outcomes. Substantial disparities in health behaviors, health outcomes, education, and social factors exist among children and adolescents of varying racial and ethnic groups. The determinants of these disparities must be better elucidated to ensure the development and implementation of effective, population-specific interventions that can be used to a) break down the financial, structural, and personal barriers to optimal health and development and b) increase access to needed care. Research is needed regarding the development and implementation of effective approaches that will enable families, health-care providers, schools, and communities to improve the health and development of children and adolescents. Effective interventions are critical to the physical and emotional well-being of children of all ages and ultimately will help young persons grow into healthy, independent, and productive members of society.

> Too many children suffer from health conditions that can be prevented. Finding better ways to promote lifelong health will give all children a chance to grow safe and strong.

1. Optimal Child Development

Develop and evaluate strategies to improve the health and fitness levels of U.S. children and to establish health behaviors that promote lifelong health by reducing the leading causes of morbidity, mortality, and disability across the lifespan.

Examples of Priority Research: Study the prevalence of risk factors, health status, and rates of disability among children. Develop broad research-to-practice initiatives to promote optimal child development (e.g., improving access to diagnosis, referral, and intervention for developmental disabilities and other adverse conditions of childhood with significant morbidity or mortality). Identify factors that promote or hinder early identification, diagnosis, and treatment of developmental disabilities and other cognitive, social, emotional, or behavioral health problems emerging in childhood and early adolescence. Evaluate coordinated developmental screening and other services for children with developmental disabilities. Assess current fitness levels among children, and examine the relationship among physical fitness and a) school policies and programs (e.g., health education, physical education, and food services); b) family knowledge, attitudes, and behaviors; and c) community-based sources of support. Translate research to increase the dissemination of effective interventions (e.g., the delivery of oral health services in school settings). Conduct economic research to identify and implement interventions that would yield the greatest health impact per dollar invested.

2. Optimal Adolescent Development

Develop and evaluate strategies to improve health and fitness levels of U.S. adolescents, and establish health behaviors that promote lifelong health and reduce the leading causes of morbidity, mortality, and disability among youth and adults.

Examples of Priority Research: Study the prevalence of risk factors, health status, and rates of disability among adolescents. Conduct dissemination research to achieve widespread adoption and implementation of proven school-, family-, faith-, and community-based interventions for promoting optimal health and development among adolescents with physical, emotional, social, and physical disabilities. Identify interventions to prevent physical inactivity, poor diet, tobacco use, alcohol use, risky sexual behaviors, and other factors that contribute to obesity, heart disease, type 2 diabetes, cancer, injury, violence, human immunodeficiency virus and other sexually transmitted diseases, and unintended pregnancy. Assess fitness levels among adolescents, and determine the relationship between physical activity and academic performance. Evaluate the impact of interventions directed to parents of adolescents that are intended to increase parental influence on risky health-related behaviors. Conduct economic research to identify and implement interventions that would yield the greatest health impact per dollar invested.

D. ADULT AND OLDER ADULT HEALTH

By 2015, the baby boomer population will represent almost one third of the U.S. population. The health challenges facing this adult population are different than those that affected past generations of adults. Because rates of overweight and obesity have more than doubled among adults during the past 30 years, this segment of the population is beginning to face the adverse health effects associated with increased weight, including type 2 diabetes, hypertension, cardiovascular disease, chronic obstructive pulmonary disease, arthritis, some cancers, and cancer survivorship. Older adults continue to face other challenges associated with the aging process. As persons age, they increase their risk for certain injuries, which can result in disability and death [38]. Research to enable individuals, families, health systems, employers, and communities to address these health challenges will greatly improve health among adults of all ages and is critical to achieving the ambitious Health Protection Goals.

Profound racial and ethnic disparities persist among adults and older adults. These disparities can be seen in health behaviors, chronic disease burden, insurance status, access to health care, health outcomes, and educational and social status. Understanding the causes of these disparities is a crucial step towards developing and implementing the most effective interventions for promoting health in all adult populations. Research to develop and disseminate the most effective adult interventions can help reduce risky behaviors that contribute to excess illness, disability, and death in adults and older adults and can enhance health and quality of life during the most productive years.

1. Reduce the Burden of, Disparities in, and Risk Factors for Chronic Diseases among Adults

Develop, implement, and evaluate strategies to establish behaviors during adulthood that promote lifelong health and reduce the risk of the leading causes of morbidity and mortality, including tobacco use, obesity, heart disease, type 2 diabetes, chronic obstructive pulmonary disease, and cancer.

Examples of Priority Research: Evaluate strategies for the dissemination and widespread adoption and implementation of proven community-based interventions that a) promote active living, healthy diets (including increased consumption of fruits and vegetables), and tobacco cessation and b) discourage the excessive use of alcohol and risky sexual behaviors. Develop, implement, and evaluate approaches to prevent or delay the onset of related chronic diseases. Evaluate interventions that impact disparities in socioeconomic status and enhance access and quality of health care, especially preventive services. Examine the burden and current and emerging risk factors for chronic conditions and comorbidities, including mental health and socioeconomic status. Evaluate the effectiveness of policy and environmental interventions. Examine the relationship between chronic disease and mental health in the adult population. Conduct economics- based methodologic research to advance health, and conduct behavioral risk surveillance.

2. Reduce the Burden of, Disparities in, and Risk Factors for Chronic Diseases among Older Adults

Develop, implement, and evaluate strategies to establish and maintain behaviors during older adulthood that sustain health, reduce the risk of chronic disease and disability, maintain quality of life, and decrease health-care costs

Examples of Priority Research: Assess the burden of and risk factors for certain chronic diseases, conditions, and comorbidities, including Alzheimer's disease, arthritis, depression, psychiatric disorders, osteoporosis, Parkinson's disease, and urinary incontinence; develop effective public health interventions to prevent or delay the onset of these conditions. Develop and evaluate interventions, including those that impact disparities in socioeconomic status, for the prevention and control of complications and disabilities that result from arthritis, cancer, cardiovascular disease, chronic obstructive pulmonary disease, diabetes, other major chronic diseases. Develop, implement, and evaluate interventions to maintain overall quality of life, including the mental and physical aspects of health. Evaluate late-life and palliative care that effectively meets the diverse needs of the growing population requiring these services.

E. HEALTH AMONG PERSONS WITH DISABILITIES

Disabilities personally and socially impact those affected and their families and caregivers. Each year, one of every 33 babies is born with a severe birth defect. Approximately 17% of children <18 years of age have a developmental disability [29], and 2% of those who are school-aged have a developmental disability defined as being severe [39]. Of all persons living in the United States, approximately 54 million are living with a disability, at an annual cost of $300 billion to the health-care system [40,41]. An estimated 24 million persons experience significant disabilities due to impairments, including cerebral palsy, rheumatoid arthritis, spina bifida, developmental disorders, inherited blood disorders, vision loss, and spinal cord injury [42]; additional challenges to the health and well-being among persons with disabilities are outlined in the Surgeon General's Call to Action to Improve Health and Wellness of Persons with Disabilities [43]. Information is lacking regarding the course of these disabling conditions and related secondary conditions (e.g., obesity and depression resulting from loss of mobility and independence) as persons move through different stages of life; the 24 million persons currently affected by severe disabilities represent the first generation of such persons to have lived into middle or old age. Understanding and preventing poor health in this population and identifying comprehensive health care practices will have a significant impact on health-care resources. Knowledge of the course and impact of illness on disabled populations will help in the development of health promotion interventions and the reduction of health disparities. In addition, research on the adverse health effects experienced by persons with disabilities, particularly those in populations most affected by health disparities, will help persons with disabilities achieve an optimal lifespan and experience the best possible quality of health in every stage of life.

> Scientific advances have helped more people with disabilities live longer lives. More information is needed to prevent disease, treat illness, and promote health so that people with disabilities can have the best health in every stage of life.

1. Health Across the Lifespan among Persons with Disabilities

Determine how illness and adverse health conditions affect persons with disabilities across the lifespan to identify promising interventions that promote health and prevent illness.

Examples of Priority Research: Study the course, prevalence, and impact of secondary physical, social, emotional, cognitive, and mental health conditions across the lifespan for persons with developmental and physical disabilities and conditions. Study the causes of congenital disabilities and chronic or hereditary conditions. Study methods to manage and minimize these problems and their effects. Determine methods of identifying and treating common chronic illnesses (e.g., heart disease, diabetes, chronic obstructive pulmonary disease, mental illnesses, and obesity) among persons with disabilities. Determine the most effective ways to promote physical fitness across the lifespan for persons with disabilities.

2. Early Identification of Developmental Disabilities

Develop and evaluate early identification and intervention strategies among children with developmental disabilities to improve academic achievement; social, emotional, and behavioral health; and physical health throughout childhood and the lifespan.

Examples of Priority Research: Design and evaluate measures for effectively tracking the development of infants, children, and adolescents with mental retardation, autism, and other developmental disabilities to identify opportunities for intervention as they emerge. Design and evaluate targeted measures to identify the specific, unique needs of children with developmental disabilities. Implement a research-to-practice initiative that identifies and implements the most effective and cost-effective interventions to maximize growth and development during childhood and adolescence.

3. Health among Infants, Children, and Adolescents with Disabilities

Determine how children and young adults with disabilities are affected by associated health and social conditions and how best to intervene to ensure their health and well-being.

Examples of Priority Research: Assess the course, prevalence, and impact of secondary physical, social, emotional, and mental-health conditions that affect children and young adults of all ages who have birth defects and developmental disabilities. Identify interventions that have the potential to maximize the health and well-being of these children, including those aimed at increasing physical activity and promoting healthy eating habits.

Chapter VI

CREATE SAFER AND HEALTHIER PLACES

People continuously interact with their environment, which can pose many preventable threats to human health, including injuries, violence, and diseases. Injuries and violence are the leading causes of death among persons aged 1–44 years and represent the leading causes of potential years of life lost [44]. In addition, work-related injuries and diseases continue to take a toll on the health of the U.S. workforce and on the economy, and environmental exposures likely are the cause of a substantial percentage of the global burden of disease. Yet, many of these diseases and injuries are preventable. Health threats arising from where people live, work, learn, and play can be reduced by applying known and effective strategies and by uncovering new, innovative ones [1,45,46].

Research to understand and address risk factors and to promote health and quality of life by preventing or controlling injuries, diseases, and deaths associated with the environment, workplace, and other settings is critical to CDC's mission. Such research will help CDC reach the Health Protection Goals for the 21st century by ensuring that the places where people live, work, learn, and play support and promote the health, safety, and overall well-being of all persons, particularly those who are at greater risk for health disparities.

To reduce the burden of environmental and occupational health threats, CDC must engage in a broad spectrum of research, including surveillance and problem identification, risk-factor analysis, intervention development, and translation and dissemination of scientific information. Accomplishing this wide range of activities requires the collaboration and coordination of researchers representing many different disciplines and organizations, including behavioral and social science, medicine, ergonomics, epidemiology, biological sciences, engineering, health education, industrial hygiene, toxicology and environmental health science, work organization, state and local public health, and a wide range of other partners. Research areas in this chapter include:

A. Environmental Health
B. Occupational Safety and Health
C. Injury and Violence

A. ENVIRONMENTAL HEALTH

The environment is everything around us—the air we breathe, the water we drink and use, and the food we consume. It is also the chemicals, radiation, microbes, and physical forces with which we come into contact. An estimated 8%–33% of the global burden of disease can be attributed to environmental exposures [47]. Populations that are particularly vulnerable to certain environmental hazards include children, the elderly, and people with disabilities. The distribution and severity of environmental exposures must be understood for different subpopulations, especially persons who are economically disadvantaged who often have the greatest risk for exposure to hazardous environmental conditions. A better understanding of the varying susceptibility of people to environmental exposures as influenced by age, nutrition, pre-existing disease, and stage of life also must be obtained, because this type of information is useful in identifying those populations at highest risk for adverse health effects and in greatest need of intervention. Examining the relationship among people, the environment, and health can help elucidate ways to prevent illness, injury, disability, and death. To maximize health impact, research focusing on the prevention of disease and injury resulting from the interaction of people and their environment should be conducted in collaboration with community stakeholders and with academic, private, and government researchers.

> Where we live is just as important to our health as how we live. Research that addresses environmental hazards will protect the quality and safety of our air, food, and water.

1. Environmental Risk Factors

Establish the major environmental causes of disease and disability, and identify related risk factors.

Examples of Priority Research: Investigate the relationship between health and the environment using a broad approach that accounts for the biological, physical, chemical, genetic, economic, and psychosocial factors that can influence health throughout all stages of life. Identify the major environmental causes of disease (including existing and emerging environmental health threats), especially among high-risk populations (e.g., children, pregnant women, and the elderly), and obtain nationally representative prevalence and incidence data. Conduct research to assess exposure to, identify risk factors for (including biomarkers of susceptibility and disease), and assess attributable risks associated with existing and emerging environmental contaminants (e.g., endocrine disruptors, aquatic toxicants, food supply contaminants, nanotechnology products, and pharmaceuticals). Conduct economic analyses to assess the burden of disease caused by environmental risk factors.

2. Complex Environmental Exposures

Develop, evaluate, and apply new and innovative methods for assessing the toxic action and health impact of multiple environmental exposures.

Examples of Priority Research: Develop and evaluate cross-cutting biomedical and modeling tools to better understand how exposure to chemical mixtures and other environmental agents impacts health. Examine the cumulative effect of multiple environmental exposures in relation to individual genetic and biological predisposition, especially among persons in vulnerable, high-risk populations.

3. Environmental Biomonitoring Methods and Tools

Evaluate the use of biomonitoring technology in humans to measure chemicals and other markers that enable exposures and health outcomes to be assessed in environmental and work settings.

Examples of Priority Research: Develop and evaluate methods and modeling tools to better understand how biomonitoring can be used to protect people from illness, poisoning (including radiation and from contaminants in food, water, air, soil and other media), birth defects, disabilities, cancer, and death caused by hazardous environmental exposures.

4. Environmental Health Interventions

Identify, develop, and evaluate the effectiveness of environmental health interventions, and elucidate best practices to prevent environmental health threats and promote health.

Examples of Priority Research: Design, implement, and evaluate environmental health interventions (including engineering controls) and health promotion activities that address complex environmental health issues (including interventions to reduce environmental risk factors associated with asthma), especially in susceptible populations (e.g., the elderly, children, persons who are disabled, and persons who are economically disadvantaged). Develop risk communication strategies, multicultural outreach initiatives, and community-based participatory research to reduce exposure to environmental hazards, and conduct intervention evaluation research that assesses behavior change in response to these public health efforts.

5. Lead Exposure and Health

Develop and evaluate policy interventions to eliminate elevated blood lead levels in the United States.

Examples of Priority Research: Design, implement, and evaluate policy interventions in communities most likely to be exposed to lead (e.g., communities with substantial numbers of racial and ethnic minorities, communities comprised of persons who have recently immigrated to the United States, and rural communities).

6. Environmental Data and Information Systems

Develop, implement, and evaluate methods and tools to link available environmental hazards and health-outcome databases to support environmental public health tracking.

Examples of Priority Research: Develop, implement, and evaluate epidemiologic, statistical, and programmatic methods and tools that can be used to link available information and further the knowledge base for topics related to environmental hazards and health outcomes. Assess and demonstrate the value of developing partnerships with organizations (e.g., Kaiser Permanente, the Veterans Administration, and Centers for Medicare and Medicaid Services) that collect health systems data to facilitate the linkage of environmental and health data regarding a wide variety of health outcomes, including chronic diseases. Develop and evaluate strategies to make data, methods, and tools for linking environmental and health data widely accessible and available to researchers from public and private sectors (e.g., academia, public health departments, and the federal government)

> Research that supports healthy workplaces keeps our workforce healthy and our economy strong

B. OCCUPATIONAL SAFETY AND HEALTH

Work-related injuries and diseases take a significant toll on human health and the U.S. economy. Each day in the United States, nearly 11,000 workers are treated in emergency departments, and approximately 200 of these workers are hospitalized [48]. An estimated 6,300 private-sector workers require time away from their jobs each day as a result of workplace injury [49], and 15 workers die from their injuries [50]. An average of 134 Americans die of work-related diseases every day [51]. Annually, these losses account for nearly $73 billion in workers' compensation claims [52] and have an overall economic impact of $155 billion [53].

Research in occupational safety and health is needed to improve the ability of employers, workers, and others to establish and maintain safe and healthful workplaces and working conditions for the 150 million civilians employed in the United States. This includes research to identify emerging hazards associated with new industrial processes; new ways of organizing work; and new chemicals, materials, machinery, and tools. Public health professionals recognize that workers typically are exposed "first and worst" to toxicants that eventually also become general environmental hazards (e.g., lead, asbestos, polychlorinated biphenyls [PCBs], and dioxin). Research also is needed to ensure that workers are better protected against longstanding health hazards (e.g., working at excessive heights, under extreme physical and psychological stress, around heavy machinery and motor vehicle traffic, in confined spaces, around excessive levels of noise, and with processes involving ionizing radiation and toxic materials and contaminants). In addition, research is needed that focuses on protecting worker populations that have high risks of occupational injury or disease as a result of susceptibilities (e.g., those associated with sex, age, disability, genetics, socioeconomics, and level of education) and on populations that are disproportionately engaged in hazardous work (e.g., low-wage workers, undocumented workers, and minority workers).

To accomplish the wide range of research activities that is needed to further understand and prevent workplace injury and disease, broad and diverse scientific disciplines must become involved. Expertise is needed from researchers who work in the fields of medicine, ergonomics, epidemiology, biological sciences, engineering, health education, industrial hygiene, intervention evaluation and translation, and work organization. Research in occupational safety and health will help accomplish the basic public health objectives of promoting and protecting the health and safety of members of the U.S. workforce; research findings will help guide the development of comprehensive programs and policies for the prevention and control of workplace-related illnesses, injuries, and ultimately deaths. This research serves a global audience and promotes U.S. efforts to improve the safety and health of all working people in every nation.

A general framework for occupational safety and health research is provided in the following text. More detailed priorities for research are being developed and will be available in the second edition of the National Occupational Research Agenda (see http://www.cdc.gov/niosh/nora/).

1. Fatal and Nonfatal Injuries at the Workplace

Identify specific risk factors associated with fatal and nonfatal injuries, and develop and evaluate interventions for reducing such injuries.

Examples of Priority Research: Improve surveillance methods to monitor health among members of the workforce and to detect and investigate health problems. Direct research efforts towards reducing health disparities among different worker subpopulations that are disproportionately affected by workplace injuries. Enhance surveillance methods and activities by exploring new sources of data, identifying ways to improve identification of work-related injuries in existing databases, linking data from various existing sources, and developing better methods to assess injury exposures and intervention outcomes. Conduct analytical research to identify injury risk factors and to quantify them by industrial sectors, priority areas (e.g., deaths from motor vehicles, machines, violence, and falls and traumatic injuries caused by falls or contact with machines, materials, equipment, and tools), and high-risk groups (e.g., construction workers, loggers, miners, farmers, farm workers, rural workers, adolescents, and workers aged >45 years). Develop and evaluate the effectiveness of interventions to prevent occupational injuries, and evaluate factors that can influence the adoption of proven technologies and strategies for protecting workers.

2. Occupational Diseases

Improve understanding of the role and burden of workplace disease exposures, and determine sector-specific risk factors, mechanisms, and effective prevention strategies.

Examples of Priority Research: Conduct research to close gaps in knowledge pertaining to the occurrence of and risk factors for occupational diseases (e.g., respiratory diseases, cardiovascular diseases, dermatitis and other skin disorders, fertility and pregnancy abnormalities, and infectious diseases). Identify high-risk occupations and exposures within industrial sectors. Evaluate workplace exposures, and develop appropriate prevention and control strategies, including modification of workplace practices, engineering controls, and protective equipment.

3. Occupational Musculoskeletal Disorders (MSDs)

Improve understanding of the relationship between ergonomic stressors and MSDs, and develop and evaluate interventions.

Examples of Priority Research: Study the causes and risk factors for MSDs, and conduct basic research to define the pathophysiologic mechanisms of chronic musculoskeletal injury. Identify reliable clinical methods to diagnose MSDs before they become severe and to rehabilitate disabled workers as fully and rapidly as possible. Identify and evaluate the effectiveness of intervention strategies intended to prevent MSDs through better cost- effective tool and equipment designs, work-rest periods, or changes to the organization of work. Develop more effective methods to promote the adoption of intervention strategies in the workplace.

4. Safe Workplace Design

Identify engineering and policy approaches that encourage the design of industrial processes that minimize occupational exposures and risks.

Examples of Priority Research: Investigate ways to enhance the prevention of work-related injuries and diseases through improvements to the designs of the physical workplace and its equipment, tools, and systems that support the elimination and reduction of hazards. Identify current and new engineering controls that help reduce injury and toxic exposures in the workplace, including the substitution of safe materials for those that are hazardous, design changes to equipment, and modification of work methods. Identify effective administrative policies for the design of workplaces and equipment that have the potential to reduce or eliminate hazards associated with occupational exposures, including modification of work practices, enhancement of management policies, and improvement of training programs. Evaluate existing and proposed engineering controls and administrative policies to determine their effectiveness.

5. Organization of Work

Determine the potential effects of work organization on workers' mental and physical health and safety, and identify intervention strategies to promote better overall health though workplace programs.

Examples of Priority Research: Conduct surveillance studies to detect and investigate health problems associated with the way work is organized. Conduct etiological research to better understand how the psychological and physical health and safety of workers is being impacted by current approaches to work organization, and determine through intervention effectiveness studies how to control adverse effects and develop and encourage the use of approaches that protect and promote the health and safety of workers. Conduct economics, industrial organization, and epidemiologic research to examine how a) work organization is influenced by the changing economy, industrial practices, and workforce demographics and b) how these changes contribute to adverse and positive health outcomes.

6. Emerging Workplace Hazards

Identify emerging hazards in the workplace resulting from changes in work practices and technologies, and develop and evaluate interventions.

Examples of Priority Research: Investigate ways to enhance prevention and to detect and investigate health and safety problems associated with emerging workplace hazards. Enhance occupational surveillance methods to facilitate identification of new hazards and their potential to cause harm to workers. Develop and evaluate control and prevention strategies as new hazards are identified. Assess the role of control technology in preventing emerging hazards, including the potential for this type of technology to a) reduce or eliminate hazards and b) be properly used in the workplace. Determine the cost-effectiveness of control technologies in responding to a new hazard.

C. INJURY AND VIOLENCE

Injuries and violence are the leading causes of death among persons aged 1–44 years, represent the leading causes of potential years of life lost, and are linked to long-term educational, occupational, physical, and mental health consequences [54-58]. For persons aged 1–34 years, unintentional injuries alone (including those resulting from motor vehicle crashes, drowning, falls, fire, poisonings, and suffocation) claim more lives than any other cause of death [59]. Fall-related injuries are the leading cause of injury deaths and disabilities among persons aged >65 years [60]. The impact of injuries resulting from all types of violence, including interpersonal violence (i.e., intimate partner violence, child maltreatment, youth violence, elder abuse, and sexual violence), is also substantial. Homicide is the second leading cause of death for persons 15–24 years of age and the third or fourth leading cause for every other age group <34 years except for persons aged 10–14 years; homicide is the fifth leading cause of death in that age group [61]. Suicide is not only the eleventh leading cause of death across all ages, but ranks second for persons 25–34 years of age and third for persons aged 15–24 years [61]. Not all violence and injury leads to death; approximately one third of all emergency department visits and 8% of all hospital stays result from non-fatal injuries [62]. Racial, ethnic, socioeconomic, and geographic factors contribute to injury disparities among populations; persons in racial and ethnic minority groups have the highest rates of unintentional injury and violence [62]. Decreasing rates of injury and violence and improving care for the injured would likely have significant and immediate effects on U.S. rates of morbidity and mortality and would improve the long-term physical, social, and behavioral functioning of persons who have experienced injuries and violence. Improvements in injury rates and post-injury care also would likely lower the cost of health care. In 2000, more than 50 million persons were medically treated for injuries in the United States, at a lifetime cost of $406 billion; $80 billion was spent for medical treatment and $326 billion for lost productivity [63].

Research devoted to the prevention of risk behaviors, the improvement of the environments most often associated with injury and violence, and the prevention of post-injury adverse health effects is needed to reduce the burden of injuries and violence all populations [64]. Examining these factors for the populations that are at greatest risk for injuries and violence (e.g., persons who are in racial and ethnic minority groups, are mentally and physically disabled, are economically disadvantaged, and live in rural communities) will help narrow existing health disparities in the United States.

1. Injury and Violence Prevention Interventions

Develop and evaluate the efficacy, effectiveness, and economic efficiency of interventions to prevent and reduce the consequences of interpersonal violence, suicidal behavior, and unintentional injury.

Examples of Priority Research: Develop and evaluate various interventions (e.g., behavioral, educational, environmental, policy, engineering, legislative, enforcement, and social/cultural) to prevent and control injuries and violence. Examine the effectiveness of programs in "real world" circumstances, and measure injury-associated outcomes, risk behaviors, and costs. Evaluate the effectiveness and efficacy of interventions using risk communications and ecological approaches.

Evaluate injury and disability outcomes in relation to costs, changes in social norms, changes in behavior, legislation, and engineering/technology improvements. Evaluate the effectiveness of interventions aimed at improving parenting and caregiver skills in preventing interpersonal violence, suicidal behavior, and unintentional injury in children and adolescents.

2. Risk and Protective Factors for Unintentional Injury

Identify the risk and protective factors associated with the leading causes of fatal and non-fatal unintentional injury in all stages of life.

Examples of Priority Research: Examine risk and protective factors associated with injury and risk-taking behavior, particularly among adolescents and other vulnerable populations that are most impacted by injury (e.g., older adults who are disproportionately affected by falls). Study the influence of peers, family, and the local social and physical environment on changing risks and risk-taking behaviors. Develop interventions that modify environments and behaviors to reduce the leading causes of unintentional injury in vulnerable populations.

3. Risk and Protective Factors for Interpersonal Violence and Suicidal Behavior

Identify the risk and protective factors associated with interpersonal violence and suicidal behavior in all stages of life.

Examples of Priority Research: Identify risk factors associated with interpersonal violence and suicidal behavior, particularly among adolescents and other vulnerable populations at increased risk for injury resulting from such behaviors. Identify protective factors that can decrease risk (e.g., strong connections to parents, family, school, religious faith, and community programs and services).

4. Trauma Systems Research

Determine and evaluate how the components of trauma systems, including disability and rehabilitation services, improve short- and long-term health outcomes and costs for the acutely injured.

Examples of Priority Research: Identify and evaluate the specific components of trauma systems (e.g., pre-hospital settings, emergency departments, and hospital-based trauma centers) and of disability and rehabilitation services that improve outcomes for the injured in both urban and rural environments. Identify the costs associated with improvements in trauma care and rehabilitation.

5. Connection among Multiple Forms of Violence

Identify the relationships among different forms of violence and other public health problems, and evaluate integrative strategies to address them.

Examples of Priority Research: Identify populations that are at increased risk for multiple forms of violence (e.g., youth violence, intimate partner violence, sexual violence, child abuse, elder abuse, and suicidal behavior). Determine the extent to which different forms of violence and other health problems (e.g., substance abuse and mental illnesses) share common risk and protective factors, and evaluate integrative primary prevention strategies to address them. Develop and evaluate primary prevention interventions designed to address these commonalities. Assess the mental-health outcomes for victims and perpetrators resulting from multiple forms of violence.

Chapter VII

WORK TOGETHER TO
BUILD A HEALTHY WORLD

The United States is committed to improving the health of people living around the world, regardless of which country they call home. CDC is working to support this effort through the development of agency-wide research priorities that focus solely on global health. Current global health research must reflect the growing need to address disease threats emerging from outside the United States as a result of increasing international travel, global commerce, and other factors that lead to the worldwide circulation of diseases and other health threats. CDC has identified several cross-cutting research areas that target a) the problems most affecting the world's population and b) those that disproportionately affect certain populations and communities including women, children, and persons who are economically disadvantaged. The research areas target major determinants of global disease and injury and focus on the tools needed for sound and effective public health action. They also address some of the greatest causes of global disease and disability and aim to diminish the substantial inequities that persist in many developing countries. International research can yield information that can be applied to health-related issues in the United States. The implementation of this research supports the three global health goals: sharing research capacity to benefit the health of global populations, furthering the understanding of external threats to better protect the health of persons living in the United States, and using research as an instrument of world diplomacy. Research areas in this chapter include:

A. Supporting Goals for Global Health
B. Disease and Injury Prevention and Control in Global Settings
C. Health of Vulnerable Populations in Global Settings

A. SUPPORTING GOALS FOR GLOBAL HEALTH

The United States and other nations around the world are committed to achieving specific goals to improve global health. These goals include those in the President's Emergency Plan for AIDS Relief [65], as well as the Millennium Declaration [66].

They reflect a U.S. commitment to address the health concerns that are most threatening the livelihood of people worldwide. Several priority concerns have been identified.

One such concern is the estimated 38.6 million people world-wide living with human immunodeficiency virus/acquired immunodeficiency syndrome (HIV/AIDS) at the end of 2005, and the estimated 4.1 million people who became newly infected that year [67]. Also concerning are the 8.3 million annual cases of tuberculosis (TB) [68] and the more than 300 million curable cases of sexually transmitted diseases (STDs) that occur each year [69], which significantly enhance HIV transmission and cause adverse pregnancy outcomes, infertility, and cervical cancer. Another priority global health problem is the millions of deaths occurring each year among children aged <5 years living in 42 countries around the world [70]; although the means to prevent these deaths are well understood (e.g., through the receipt of vaccine and consumption of micronutrients), preventive measures remain unavailable to millions of affected children. Other priority health concerns include a) the major disparities in maternal mortality that exist from country to country and b) the lack of safe drinking water in many countries around the globe; each year, 1.6 million deaths (most of which occur among children <5 years of age) can be attributed to unsafe water and to poor sanitation and hygiene [70].

The achievement of these global health commitments will impact the health of persons around the world by reducing major health inequities. To be effective, however, the interventions needed to achieve global health goals in diverse settings must be guided by operational research. Such research will benefit large populations of persons living in countries of greatest need and will help achieve the following global health goals: a) reducing child mortality; b) increasing life expectancy; and c) decreasing disease and disability among persons of all ages.

> Poor nutrition and disease prey on the world's most vulnerable, including the millions of children who die each year before their fifth birthdays. By reducing the spread of HIV, increasing childhood immunization, and improving nutrition, we can help more children grow safe and strong

1. Global Mortality among Mothers and Young Children

Develop, implement, and evaluate interventions that increase maternal and child survival in priority countries to meet the respective goals of the Millennium Declaration [66].

Examples of Priority Research: Conduct operational and evaluation research to obtain essential information needed to guide the development and implementation of programs to prevent deaths among mothers and young children. Priority topics include a) emergency obstetric care; b) the practice of spacing and limiting births; c) breastfeeding practices; d) diseases that affect newborns; e) the major causes of child mortality (e.g., malaria, diarrheal diseases, pneumonia, and measles); f) the major risk factors for mortality (e.g., malnutrition); g) critical areas of intervention (e.g., family planning and appropriate drug therapy); and h) health-system needs for providing effective interventions.

2. Immunization to Eliminate and Protect against Global Diseases

Evaluate interventions to increase the impact of global immunization programs.

Examples of Priority Research: Determine the field effectiveness of current and new vaccines, and assess the effectiveness and economics of different strategies to increase immunization coverage. Evaluate the integration of immunization with other health interventions. Identify new technologies for vaccine-preventable disease surveillance in resource-poor international settings, including developing field-appropriate methods for the detection and laboratory confirmation of measles, rubella, poliomyelitis, and other vaccine-preventable diseases. Evaluate the morbidity, mortality, and economic impact of diseases for which established vaccines exist or new vaccines are being developed. Evaluate disease elimination efforts and strategies, including evaluation of national immunization programs, policy, surveillance, and other strategies. Evaluate the barriers to immunizing children according to the current recommendations in both developed and developing countries.

3. Global Micronutrient Malnutrition

Develop, implement, and evaluate interventions to prevent vitamin and mineral deficiencies of global importance.

Examples of Priority Research: Improve micronutrient (e.g., iodine, iron, zinc, vitamin A, and folate) assessment methods. Investigate the efficiency of different micronutrient delivery strategies, including fortification of foods, and evaluate program effectiveness in priority settings. Develop methods to quantify the problem of micronutrient deficiencies in stable and complex humanitarian emergency environments, and identify interventions to treat and prevent such deficiencies.

4. Global Efforts Regarding Human Immunodeficiency Virus/Acquired Immunodeficiency Syndrome (HIV/AIDS), Tuberculosis (TB), and Sexually Transmitted Diseases (STDs)

Develop, implement, and evaluate approaches regarding the prevention and treatment of, diagnostic testing and surveillance for, and delivery of services associated with HIV/AIDS, TB, and STDs.

Examples of Priority Research: Develop, implement, and evaluate behavioral and biomedical interventions to prevent and treat HIV infection among adults and children and to integrate reproductive services with HIV prevention. Identify behavioral interventions to promote knowledge of HIV status and treatment adherence. Identify simpler, less costly diagnostic techniques. Develop and evaluate interventions to prevent or reduce STDs among vulnerable populations. Conduct operational research to strengthen global TB surveillance and control. Develop and evaluate new and more rapid diagnostic testing for multidrug-resistant TB.

5. Global Water Safety

Determine the burden of illness associated with water supplies from untreated sources in global settings, and investigate interventions for the prevention of diseases.

Examples of Priority Research: Determine the burden of illness associated with water supplies from untreated sources, and identify existing risks to vulnerable populations (e.g., young children and people living with HIV/AIDS). Assess the efficacy and effectiveness of interventions that have a high likelihood of being sustainable in international settings (e.g., water treatment at the household level). Identify barriers to access for appropriate technologies, evaluate the efficiency and effectiveness of innovative implementation strategies in increasing access to safe water in target populations, determine the cost-effectiveness of different appropriate technologies and implementation approaches, and assess the effectiveness of different behavior-change techniques in motivating adoption of appropriate technologies. Conduct operations research to identify factors that will enhance the successful implementation of large-scale programs.

> Disease, injury, and violence threaten the health of people in all countries. CDC and its partners can work together to promote healthy people in a healthy world.

B. DISEASE AND INJURY PREVENTION AND CONTROL IN GLOBAL SETTINGS

The major causes of morbidity and mortality in the world must be elucidated before disease and injury prevention and control efforts can be successful in global settings. Of the approximately 57 million deaths worldwide in 2003, an estimated 15 million (>25%) were caused by infectious diseases [71]. Although a subset of these diseases has been targeted for elimination, new infections continue to emerge, and existing diseases re-emerge in more virulent forms, making prevention and control efforts challenging. In 2003, 46% of the global burden of disease and 59% of the 57 million deaths were caused by non-communicable chronic conditions, including cardiovascular diseases, diabetes, obesity, cancer, and respiratory diseases [72].

More than 5 million people around the world die each year from injuries, with 1.2 million deaths being traffic-related [73,74] and 1.6 million deaths resulting from acts of violence [54]. The rates of violent death in low- to middle-income countries are more than twice as high (32.1 per 100,000) as those in high-income countries (14.4 per 100,000) [54]; in addition, most drowning deaths, traffic-related deaths, and pedestrian deaths occur among children in low- and middle-income countries [74,75]. Substantial differences in the epidemiologic patterns and the rates of death and disability from injuries and violence can be observed among world regions and across age groups. Many countries do not have the capacity to provide prompt emergency care, including rapid movement of injured victims from the scene of injury to a health-care facility [76], which likely contributes to global disparities in health outcomes.

Globally, 1.7 million deaths each year are caused by illness and injury associated with workplace hazards. Each year, 268 million non-fatal injuries and 160 million new cases of work-related illness are reported [77].

Public health interventions directed at the prevention and control of the diseases and injuries that cause the most morbidity and mortality are among the most effective ways to improve the health of large populations. Such public health interventions identify those populations most in need and can be employed to reduce disparities by better reaching groups that do not have access to traditional modes of care.

1. Global Infectious Diseases

Develop and evaluate functional tools to detect, diagnose, predict, and respond to endemic and emerging infectious diseases of global or regional importance.

Examples of Priority Research: Develop and assess new tools (e.g., drugs, diagnostics, and vaccines), surveillance methods, treatment regimens, and intervention strategies for the control, prevention, and eventual elimination of infectious diseases. Address the following priority research areas: a) infectious diseases that pose the greatest threat to public health; b) infectious diseases targeted for elimination; c) infectious diseases associated with high fatality rates; and d) infectious diseases with substantial potential for international transmission. Assess the burden of infectious disease caused by specific pathogens to assist decision-making regarding the introduction of new vaccines.

2. Global Burden of Non-communicable Diseases

Define the magnitude of global non-communicable diseases, identify the major risk factors for the disproportionately high burden of such disease in the developing world, and develop and implement interventions for disease prevention and control.

Examples of Priority Research: Assess methods for quantification of the burden of non- communicable diseases in developing countries. Identify the individual and contextual determinants that shape health behaviors of individuals, groups, and communities, as well as the environmental determinants contributing to such disease. Develop new intervention strategies to reduce the risk factors associated with unhealthy behaviors. Priority research areas include cardiovascular disease, hypertension, cancer, diabetes, mental illness, and oral health, as well as major risk factors (e.g., tobacco and obesity).

3. Global Burden of Injuries

Identify the causes, consequences, and costs of intentional and unintentional injuries, and assess the efficacy, effectiveness, and economic efficiency of interventions to prevent injuries and mitigate consequences.

Examples of Priority Research: Determine the relationship among various aspects of globalization (e.g., economic, environmental, and cultural factors) and violence and injury (both intentional and unintentional). Identify risk and protective factors common to different population groups, cultures, and settings. Develop primary and secondary prevention approaches that impact all populations in various settings, focusing on developing countries. Evaluate diverse international public health programs created to prevent injuries, improve injury care, and reduce adverse health outcomes.

4. Global Occupational Health

Identify, develop, and evaluate effective interventions to reduce hazardous exposures and safety risks at the workplace that contribute to the high rates of work-related disease and injury in developing nations.

Examples of Priority Research: Evaluate interventions from developed countries to determine whether they can be adapted for use in both urban and rural work settings in developing countries. In conjunction with existing preventive measures and in light of local operational constraints, develop technologies to ensure the continuing health and well-being of persons in the workplace. Inventory, compare, and contrast intervention alternatives relative to health effects, costs, and benefits. Identify and develop user- friendly model(s) to determine the type of intervention required. Evaluate strategies to build capacity and professional collaborations in developing nations to overcome barriers and satisfy global workplace needs.

C. HEALTH OF VULNERABLE POPULATIONS IN GLOBAL SETTINGS

The current health status of vulnerable populations around the world (e.g., refugees and internally displaced people, exploited women and children, orphans, and people with disabilities) is poorly understood, primarily because these populations are largely marginalized. An estimated 45 million people are affected by complex humanitarian emergencies (CHEs) resulting from war, natural disasters, and other catastrophes [78]. Most causes of death and disability among persons affected by CHEs (e.g., measles, malaria, diarrhea, acute respiratory infections, threats to maternal and neonatal health, injuries sustained from land mines, and pedestrian- and cycle-related injuries) can be prevented with known interventions. In 2002, the International Labor Organization estimated that 250 million persons aged 5–17 years were involved in child labor; approximately 179 million of these children routinely engaged in severe forms of such labor (i.e., those that endangered their physical, mental, or moral well-being) (79). According to the Standing Against Global Exploitation (SAGE) Project, an estimated 10 million women and children are exploited through the commercial sex industry each year, which places them at risk for human immunodeficiency virus/acquired immunodeficiency syndrome (HIV/AIDS) and other sexually transmitted diseases (STDs) as well as physical and psychological abuse [80].

The United Nations Children's Fund (UNICEF) estimates that by 2010, 106 million children <15 years of age will lose one or both parents because of illness, violence, or injury; an estimated 25 million of these children will be orphaned as a result of HIV/AIDS alone [81]. Orphans and other vulnerable children are often left unprotected after loss of parents, placement in temporary shelters, or loss of contact with caregivers. They are often more vulnerable to becoming victims of violence, exploitation, human trafficking, discrimination, and other abuses. In times of conflict (e.g., war), these children experience increased risk of exposure to violence, physical abuse, and exploitation, along with an increased risk for death. These vulnerable children face disease, malnutrition, physical and psychosocial trauma, and impaired cognitive and emotional development. Girls are at especially high risk for sexual abuse, whereas boys are at high risk for being conscripted into conflict as child soldiers.

Descriptive epidemiologic research focused on the most effective interventions for decreasing health disparities in vulnerable populations is needed. This research should a) identify interventions that can be used to modify the determinants and the progression of health problems associated with the highest disease burden and b) promote disease prevention among vulnerable populations in global settings. Achieving a broad impact and reducing disease burden worldwide will require not only medical interventions, but also behavioral, social, and political interventions implemented by sectors other than those that traditionally focus on health.

> War, disease, and natural disasters threaten the lives of millions of women and children around the world. Global efforts to prevent disease and protect health can make the world a healthier, safer place for all women and children.

1. International Complex Humanitarian Emergencies (CHEs)

Assess the key risk factors and the disease and injury outcomes associated with CHEs, and identify public health prevention measures to decrease illness and death.

Examples of Priority Research: Rapidly assess the disease, injury, acute care, and disability burdens associated with CHEs (e.g., war, societal conflict, natural disasters, and displacement of populations) to help allocate resources toward resolving major health issues. Develop, standardize, and implement reproductive health indicators, interventions, and services in emergency and post-emergency settings. Assess local, national, and international response capacities and resources as a proxy for sustainability. Improve training methods for the analysis and interpretation of indicators to target interventions. Develop, test, and evaluate standardized health information systems in emergency and post-emergency situations. Develop new research methods to quantify the problem of micronutrient deficiencies among populations involved in CHEs, and identify interventions to prevent such deficiencies and treat persons who are impacted by them. Examine the mental health needs and coping mechanisms of persons of all ages involved in CHEs using cross-cutting tools that include ethnographic and sociological research methods. Determine the long-term physical, social, and behavioral functioning of persons experiencing CHEs.

2. Public Health Consequences of Exploitation of Women and Children in Global Settings

Examine the root causes of exploitation and the global interventions needed to prevent it.

Examples of Priority Research: Determine the sociopolitical factors and descriptive epidemiology of exploitation, and identify the magnitude and scope of exploitation- related diseases. Examine the prevalence and trends of sexual exploitation, impact of public policy, role of the Internet in proliferation, resulting psychosocial and medical effects, cultural variations, and role of prevention education. Assess the impact and extent of sexual violence and sexual exploitation on women and children during armed conflict and other CHEs situations. Identify the unique health problems of women and children who are victims of exploitation (including child labor), and investigate the types of services most needed by these populations.

3. Orphans and Other Vulnerable Children in Global Settings

Develop and evaluate interventions to reduce the adverse psychological and physical health problems of orphans and other vulnerable children.

Examples of Priority Research: Assess interventions to improve family and community capacity to care for orphans by reducing stigma and strengthening support systems. Describe the incidence and prevalence of children with developmental delays, psychosocial conditions, disabilities, and diseases. Assess the impact of armed conflict and other CHE situations on children and other vulnerable populations.

D. SOCIETAL DETERMINANTS OF HEALTH IN GLOBAL SETTINGS

People who are socially and economically advantaged generally experience better health than those who are disadvantaged, regardless of whether they live in developing or industrialized countries (82-88). To help reduce global health disparities, the relationship between socioeconomic status and health at the individual and community level must be better elucidated. An adequate understanding of the social and economic factors that affect health in various settings is the first step towards narrowing social inequalities in health. At the same time, understanding the mechanisms through which socioeconomic circumstances influence individual and community health is likely to create a knowledge- base for effective and efficient interventions. For instance, understanding whether individual or community socioeconomic status and health are linked through a) differential access to material resources (e.g., clean water, adequate housing conditions, and access to health care) or b) differential psychosocial influences associated with socioeconomic status (which impact stress-related physiological systems) may help elucidate the causes of health disparities at the individual and community level (89). The value in viewing individual health outcomes and individual risk factors within their proximate and distal context (e.g., neighborhoods and political units [such as states]) is increasingly being recognized. In addition, contextual factors (e.g., income inequality, social capital, and human capital) have been suggested as being potential societal risk factors that impact health (84-88). Operational research within a multilevel causal framework (89-91) is therefore needed to determine and assess the impact of specific social and economic circumstances and interventions on health, injury, and disability on a global scale.

> Public health research can uncover the social and economic factors that influence health and find better ways to reduce health disparities around the globe.

1. Relationship between Socioeconomic Status and Global Health

Evaluate the linkages between social, cultural, and economic status and health outcomes throughout the world.

Examples of Priority Research: Describe associations between different socioeconomic indicators of health (e.g., education, income, social status, and race/ethnicity) within a system framework and across different stages of life. Evaluate the linkages between contextual factors measured at different societal levels (e.g., social capital and income inequality) and health, both within and between nation states. Examine the pathways through which contextual factors influence health, as measured through mortality, disease, behaviors, and physiological markers.

2. Human Resources and Health Outcomes in Global Settings

Evaluate the way the quality and quantity of human resources affects the key determinants of health outcomes in global settings.

Examples of Priority Research: Evaluate the linkages among health-care associated human resources and health outcomes in various local, regional, and national level settings to set resource allocation and training priorities at these levels. Examine whether the available human resources are differentially utilized by each section of the population. (a) Assess the distribution of limited human resources (e.g., provision of healthcare in rural settings); most appropriate types of human resources in financially limited settings; c) quality of services provided; d) supervision, training, and education of health-care workers; and e) involvement of traditional healers and other non-formal health-care workers in ensuring the health of a population. Examine how working with communities and non-traditional partners can lead to alternative solutions for delivering essential health interventions.

E. TOOLS FOR GLOBAL PUBLIC HEALTH

The need for global public health action is well recognized. However, substantial variation exists in the public health infrastructure of the major regions of the world. Many countries face challenges in taking action to improve health because they lack a) appropriate, well-evaluated, and researched interventions and marketing strategies and b) thorough assessments of the scope of region-specific public health problems. Several sources of data on diseases, injuries, and health-related variables currently are available for many regions of the world; however data quality varies greatly among countries, and inconsistencies exist in the way indicators are measured. Lack of available data is particularly apparent in the fields of non-communicable diseases and injuries. To positively impact health outcomes around the world, the comparability of health data (i.e., information on disease, injury, death, and disability) must be improved across countries, and the contextual nature of many variables of interest (particularly personal risk factors and social determinants) must be taken into account [85,89-91]. Although it is a recent undertaking in the United States and other western countries, social marketing based on a sound theoretical and practical approach is widely perceived as being an effective way to improve a population's health. However, research regarding the delivery of health messages to the public in most developing countries has been sparse. A critical need exists for evaluation of public health messages and the role they play in improving global health. Improving global strategies for public health action can lead to the reduction and prevention of injuries, disease, and disability and help reduce health disparities among countries and populations. The following three areas of research can have broad impact on global health: a) standardizing health measures; b) improving strategies for distributing health products and methods; and c) assessing global intervention effectiveness, including cost-effectiveness.

1. Global Measurement of Health, Disease, and Injury

Further develop and enhance ways to systematically measure health, disease, and injury outcomes that allow for within- and cross-country comparisons.

Examples of Priority Research: Determine appropriate outcome measures for within- and cross-country comparisons. Develop and evaluate harmonized survey instruments and methods for collecting data across countries. Determine the appropriate application of forecasting as a research tool for data analysis. Further elucidate the sociocultural factors that play a role in data comparability issues. Investigate state-of-the-art analytical procedures that can be used to create consistencies in data comparison across and within countries. Research the state of surveillance and infrastructure.

2. Health Marketing and Health Education in Global Settings

Improve strategies for health education and dissemination of health marketing messages and products with the objective of encouraging behaviors that will prevent disease and injury in the developing world.

Examples of Priority Research: Determine best practices on how to adopt existing marketing approaches used in commercial industries and apply them to address health issues in developing countries. Determine how to deliver health messages (e.g., through community-based institutions and social entrepreneurs), and assess the effectiveness of communications aimed at changing behaviors. Identify cost-efficient marketing strategies for distributing and encouraging the use of health-promoting products (e.g., vaccines, micronutrients, condoms, bed nets, cycle helmets, smoke alarms, life preservers, preventive medications, and treatments to purify home drinking water treatment) within the sociocultural context of developing countries. Evaluate strategies for improving health literacy in the global context.

3. Evaluation of the Effectiveness of Global Health Interventions

Assess the impact of different interventions used to address priority global health problems in developing countries.

Examples of Priority Research: Develop a knowledge-base of effective interventions, including those aimed at reducing health disparities. Assess the cost-effectiveness for health interventions that address the substantial causes of morbidity and mortality in the developing world. Assess contextual, social, cultural, and economic factors that interfere with the coverage or effective delivery of existing health interventions.

Chapter VIII

MANAGE AND MARKET
HEALTH INFORMATION

The public health needs for managing and marketing health information and for developing the tools, techniques, and strategies for this work have never been greater. Health statistics, public health informatics, and health marketing are essential foundational and cross-cutting areas that allow public health professionals to improve and manage health information. These disciplines enable the tracking and monitoring of information and help public health professionals reach and engage individuals, communities, and entire societies. Each of these areas is rooted in long-standing, science- based disciplines that make strong and unique contributions to health protection, prevention, and promotion. Research in these disciplines must be expanded, and collaboration among public health entities and organizations must be increased. Such expansion will enable public health professionals to have a greater impact on each of the priority health areas discussed in the Research Guide. Additional research also is needed to address issues associated with the *Health Insurance Portability and Accountability Act* [8] and confidentiality issues in public health data management and surveillance.

Public health data serve as the basis for guidelines intended to help people live longer and healthier lives. The most accurate and reliable data are collected, managed, and analyzed by researchers, public-health practitioners, and health-care professionals to further identify factors that have an adverse impact on health and well-being and to develop interventions that will improve the health of all persons, particularly those affected by health disparities. The field of public health informatics plays a critical role in integrating the overall strength and importance of disease surveillance data. The development of this field and dissemination of informatics knowledge and expertise to public health professionals is the key to unlocking the potential of information systems to improve the health of the world.

Health marketing is a strategic, integrated approach to health protection that involves creating, communicating, and delivering science-based health information to meet the needs of diverse customers and stakeholders, including individuals, organizations, communities, and populations. Combining the science and best-practices of marketing with the core values of public health addresses the growing need of the public health community to reach intended audiences more quickly and with more targeted and tailored messages. Research in health marketing should foster new collaborations and intellectual exchange of public health information across diverse disciplines and research communities. Research categories in this chapter include:

A. Public Health Data
B. Public Health Informatics
C. Health Marketing

A. PUBLIC HEALTH DATA

The practice of public health involves collecting, analyzing, and sharing data that drive evidence-based decisions with the goal of improving health impact. Without robust data, public health researchers and other allied health-care professionals cannot successfully answer public health research questions. Public health data provide the backbone for understanding different public health research hypotheses. One of the biggest concerns facing public health is the need to effectively manage the public health data currently being collected. Innovative data management strategies, along with statistical methodologies, are being created to manage and mine public health data for use in a wide range of public health disciplines. In addition, research currently is being conducted that will allow for the development of a formal system in which surveillance and epidemiologic data can be shared across a number of federal agencies. Strategies and methodologies also are being developed to forge a bridge between public health data and health-care data collected in private clinics, hospitals, and insurance companies [92].

With the effective management and mining of public health data, gaps in knowledge regarding health disparities are more easily identified and addressed. Public health data have the potential for broad impact, because such information is useful to scientists and researchers across many health disciplines. Public health data are relevant to all of the Health Protection Goals developed by CDC and support the achievement of improved health and quality of life. Research that merges the practical needs of the field and the scientific rigor that CDC provides can help to develop new and better approaches to the effective use of data for public health decision-making and practice.

> Better health data means better decisionmaking—and that means better health for all Americans.

1. Statistical and Data Science

Identify, develop, and provide new and innovative statistical, demographic, and epidemiologic methods for improving the way data are used to address public health issues.

Examples of Priority Research: Conduct basic research in statistics and related fields to address limitations of current methodologies and approaches in the statistical use of data to address public health issues.

2. Data Collection

Identify and develop efficient and effective data collection methodologies to ensure that accurate, timely data are available for analysis and interpretation by public health practitioners, health-care professionals, and professionals in other disciplines.

Examples of Priority Research: Conduct research to develop data collection methods and procedures that capture a wide range of public health problems and related factors (e.g., innovative sampling strategies for obtaining information for population subgroups; innovative study designs [including longitudinal and cohort studies] that can be used to evaluate interventions; reliable and valid data collection methods [including questionnaires and data capture forms]; and new methods for obtaining data [including web-based modalities]). Investigate which data items are the most essential to the attainment of a range of objectives, including those associated with emerging issues (e.g., indicators that signal events associated with notifiable diseases).

3. Data Integration

Identify and explore ways to integrate data from different sources.

Examples of Priority Research: Conduct research to develop a surveillance system that links clinical and public health information. Conduct research that leads to the establishment of a foundation for direct data linkage with public health surveillance systems and other clinical-care data systems. Conduct research that helps elucidate the barriers to the integration of public health data and other health-related data systems, and identify ways to overcome those barriers. Investigate strategies for using integrated data to inform policy decisions.

4. Data Analysis

Identify and develop new and innovative analytic techniques to address public health issues.

Examples of Priority Research: Conduct research to develop analytic techniques that maximize the use of a wide range of data, including methods that address the analysis of data from multiple sources, the use of longitudinal data, and the use of data collected for other purposes. Investigate methods that allow users to perform automated and real-time analyses of data to identify patterns of interest, process data in free-text form, and query large data bases to test existing databases or generate new ones. Conduct research into how to protect confidentiality while promoting utility.

5. Data Dissemination

Develop ways to maximize access to data and the dissemination of results to a range of users with different needs.

Examples of Priority Research: Conduct research to develop innovative methods that allow users to access public health data using various approaches while protecting privacy and confidentiality. Investigate ways to make data understandable and available to users with different needs and technical abilities.

B. PUBLIC HEALTH INFORMATICS

Public health informatics is the systematic application and integration of information (e.g., computer science and other technology) into public health practice, research, and learning [94]. Informatics and information systems are critical to the practice of public health in the 21st century. Fully operational information systems must be put into place to address emerging and routine public health needs among various U.S. populations and to help elucidate factors that contribute to health disparities in certain groups. These systems will provide new and creative solutions to extend the reach of public health, allowing it to achieve more. The need for informatics development has been recognized within all public health disciplines.

Public health informatics can provide new, high-impact capabilities for preventing and managing diseases and other public health threats. Within the U.S. public health system, informatics as a public health discipline is increasingly being recognized as a focus area capable of improving all aspects of public health [95]. Informatics-related research has the potential for broad impact by helping public health professionals define and manage the architecture for national public health information systems [95]. This kind of research identifies the capabilities that are needed to ensure that these systems work together and connect with clinical care and other health organizations. Also needed is research to develop a scientifically grounded understanding of the information needs of the public health workforce, during both emergency and routine decision- making. Research to develop methods for collecting, analyzing, and assessing the public health literature and for defining the evidence base in public health knowledge also is needed, along with research to a) develop public health ontologies and vocabularies that consistently structure public health datasets and information and that can enable knowledge sharing; b) design and evaluate models to identify, filter, organize, synthesize, and disseminate evidence-based public health information both nationally and internationally; c) design and evaluate models to intelligently and transparently link multiple components of the public health knowledge-base to public health practice; and d) identify and quantify training needs of the public health workforce relative to information use and informatics.

Public health informatics can be used to reduce health disparities, because this discipline allows public health professionals to rapidly and consistently identify gaps in public health needs. Public health informatics could have a broad impact on the health of all persons in the United States, because it can be used across all public health disciplines. Its relevance to all of the Health Protection Goals developed by CDC stems from research methodologies and strategies that focus on informatics in public health research.

1. Analytical Methods for Informatics

Develop algorithms and conduct analyses for public health detection and for monitoring diseases and health conditions.

Examples of Priority Research: Investigate ways to enhance outbreak detection and disease and health-condition monitoring through advancements in pattern recognition, algorithm development, and analytical methods. Develop statistical surveillance approaches, algorithms, and processes for the identification of health events and trends. Enhance the public health algorithms and approaches needed to satisfy reportable disease and other public health data needs, which will provide the foundation for disease detection and monitoring strategies. Explore the use of qualitative research methods in public health investigations.

2. Information and Data Visualization

Explore approaches and best practices for visualizing, analyzing, and mapping public health data.

Examples of Priority Research: Investigate ways to visualize information and data to enable a) greater focus on spatial and geographic relationships and context and b) small area analysis (e.g., census tracts and individual housing blocks). Examine dynamic mapping practices, geographic information systems, analysis of spatial data, environmental modeling, and plume modeling for air-dispersal patterns. Develop approaches for collecting population and other public health data and for creating probabilistic population-based representations of communities. Create better information and data visualization tools that will lead to more real-time monitoring of systems by providing a) the current status of and reports from each system and b) snapshots at regular intervals for trend analysis.

3. Communications and Alerting Technologies

Explore strategies and technologies to facilitate communication and collaboration among the diverse sectors focused on health protection activities.

Examples of Priority Research: Investigate technologies and practices to enable public health professionals to communicate information and data to the appropriate persons at the appropriate time and in the appropriate format. Investigate how to most effectively apply information technology to support the efforts of federal, state, and local governmental public health agencies. Examine the importance of coordinating "virtual team" technologies that enable geographically disparate groups to work efficiently together and share information during an emergency or crisis. Evaluate cost-effectiveness and dissemination practices of communication strategies and alerting techniques.

4. Decision Support

Explore technologies and methodologies that will assist decision-makers in creating public health solutions.

Examples of Priority Research: Identify technologies and practices that can a) assist key decision-makers, call-response personnel, public health practitioners, and subject matter experts in determining the most probable cause of a problem and b) facilitate the creation of decision trees that outline ways to solve these problems in accurate and scientific ways. Conduct research that facilitates the development of a detailed series of recommendations drawn from robust knowledge, management repositories, decision trees, adaptive learning tools, and fault modeling technologies. Employ consequence management technologies to evaluate the potential impact and outcomes associated with public health responses and to help key decision-makers make rapid and informed determination of risks. Evaluate approaches for the presentation of community health event trends for health-care providers. Examine approaches for evidence-based public health decision support systems and prioritization systems. Develop and evaluate public health expert systems and tools (i.e., those based on artificial, or computerized, intelligence) to assist key decision-makers both within and outside of health-care environments. Develop and evaluate tools to assist public health professionals in making rapid and informed determinations of risk.

Improve measurement tools and data evaluation strategies regarding public health outcomes.

5. Electronic Health Records

Explore practices and strategies for creating and using electronic and personal health records for public health.

Examples of Priority Research: Better inform the decision-making and practice of clinical- care professionals by developing and evaluating strategies that increase the use of public health guidelines and by ensuring that they receive updated public health information on a variety of topics (e.g., specific diseases, health conditions, and effective interventions). Identify workflow tools for converting manual, paper, and/or redundant processes to electronic and automated, integrated systems. Develop and improve methods for integrating public health functions into electronic health records. Investigate approaches for the integration of local, regional, and state outbreak management and public health decision support. Explore multivariate aggregation and analysis from disparate clinical systems. Evaluate health-care data source availability, data translation, and data communication methods for public health use. Examine ways to protect the security of electronic health data. Integrate public health functions and needs into the electronic health-care environment while fostering public health informatics solutions that can be used in other local and regional settings. Develop methods for integrating public health functions into telemedicine and electronic health record systems. Investigate approaches for the integration of local and regional data into state and national systems to improve outbreak management, public health decision support, public health communications and alerting, and countermeasure and response administration capabilities. Explore approaches regarding multivariate aggregation and analysis of data from disparate health-care systems. Research ways to create robust, standardized, and accurate health information.

6. Knowledge Management

Explore strategies and approaches for efficiently managing public health knowledge so that it can be appropriately used according to need and audience.

Examples of Priority Research: Investigate practices and technologies that could render public health knowledge more accessible (e.g., capturing successes and lessons learned, storing and classifying information to facilitate quick retrieval, and repurposing content).

Research and evaluate public health cultures of information and data sharing that can be used to build stronger collaborative health networks. Explore technologies, methods, and approaches to support both ad hoc and long-term collaboration. Evaluate and develop flexible public health expertise systems that connect extant and surface nascent public health networks to optimize information and knowledge sharing and dissemination. Explore and develop rich public health vocabularies and ontologies to support the formation of relevant associations between information contained in distributed data and information repositories.

Research and develop processes to provide increasingly relevant public health knowledge in the context of the personal health record and evaluate its usefulness. Research and evaluate the public health literature to define the evidence-base in public health knowledge. Integrate knowledge retrieval and presentation methods and algorithms (e.g., situational awareness, data visualization, and decision support) into other public health systems (e.g., surveillance, outbreak management, and electronic/ personal health record systems) to facilitate robust knowledge awareness.

C. HEALTH MARKETING

Health marketing is a dynamic and emerging area of public health that allows agencies and organizations to more effectively and efficiently exchange public health information with relevant audiences. It is a cross-cutting, trans-disciplinary approach to health protection that involves creating, communicating, and delivering science-based health information to meet the needs of diverse customers and stakeholders, including individuals, organizations, communities, and populations [96].

Health marketing research is needed to identify methods for rapidly reaching intended audiences and to help public health professionals better understand audiences, which helps ensure more appropriate and timely information dissemination. Research in health marketing also helps to reduce health disparities by identifying and emphasizing the unique qualities and needs of all public health customers, especially those at greatest risk for health disparities and adverse health outcomes. Innovative health marketing research is needed to increase the applied integration of recent advances in numerous diverse disciplines (e.g., marketing, health communication, social sciences, neurobiology, informatics, business theory, information theory, mathematics, and evidence-based medicine) to ensure that the most effective, appropriate, and ethical practices are used to create, communicate, and deliver health information and to improve and protect health throughout the United States and around the world.

Research and practice in health communication is particularly critical to the success of health marketing initiatives, in that the science, theories, principles, and strategies of health communication inform the development and dissemination of targeted and tailored health messages to diverse intended audiences.

Research is needed to further explore how health communication can be leveraged to increase public health's impact across many topic areas and to ensure that the public has the information needed to make informed health decisions. Health marketing research can have a broad impact; the knowledge gained from this type of research is applicable to almost all fields of public health. Health marketing research has great relevance to CDC's overall mission—it cuts across all stages of life, settings, and populations, and can be used to reach health protection and promotion goals in the United States and worldwide.

1. Informed Consumer Health Choices

Develop, evaluate, and implement strategies and approaches for consumers that enable them to evaluate and self-manage their health.

Examples of Priority Research: Develop feedback and measurement tools and techniques for consumers to enable them to assess their own health. Develop effective approaches for presenting health information across racial and ethnic populations using a variety of communication media. Investigate ways in which various models of health (e.g., the medical model, public health model, wellness model, and community health model), as perceived and accepted by individuals and groups, impact health-seeking behavior and response to public health messages.

Explore the impact of certain behavior models (e.g., the addiction model) on individual and group acceptance of public health interventions. Identify the practical implications (e.g., economic and social) of incorporating specific models of health into health marketing campaigns.

2. Integrated Health Marketing Programs

Explore practices and strategies to enable public health professionals to design, create, and evaluate effective and integrated health marketing campaigns.

Examples of Priority Research: Develop and test approaches and strategies that will provide health professionals with the information and tools necessary for creating integrated health marketing programs and public health initiatives. Generate new knowledge regarding the process of changing social norms, expectations, and public decision-making protocols to create healthier environments and empower individuals and groups to succeed in changing negative attitudes, activities, and behaviors. Develop and study the role of varying definitions and measures of "success" in health marketing for different partners and audience groups (e.g., consumers, caregivers, health-care and public-health practitioners, policymakers, employers, and other stakeholders). Promote research that facilitates the identification of strategies to promote knowledge transfer and uptake among public health decision-makers, policymakers, and practitioners. Investigate ways to translate this information to achieve greater success in public health programs and interventions. Develop and evaluate alternative marketing and dissemination strategies for improving population health among rural, minority, economically disadvantaged, and other underserved populations.

3. Health Awareness to Health Action

Identify and develop strategies that will equip those who develop health marketing campaigns to move individuals from health awareness to health action, resulting in a more sustained health impact.

Examples of Priority Research: Determine how the design, delivery, and reception of messages contribute to informing and instigating behavior change in various populations. Develop and evaluate health messages, and address factors for sustaining health behaviors through successful communication interventions. Generate new knowledge regarding the process of inducing voluntary changes in a) the attitudes and behaviors of persons regarding their own health and b) activities and approaches of key societal groups that impact the health of defined populations. Elucidate the theory and socioeconomic behaviors that lead to rapid acceptance of, and motivation for, change regarding health and health behaviors. Evaluate the role of regulation and the impact of various social drivers on acceptance of specified health behaviors (e.g., whether smoking bans increase the perceived value of specified components of health, such as respiratory and cardiovascular health, or of related interventions, such as nicotine patches).

4. Niche Marketing

Identify measurable, accessible, sensitive, and sustainable target audiences to create highly targeted health messages for urgent health impact.

Examples of Priority Research: Study niche marketing and its relevance to public health interventions. Investigate what level of market differentiation is necessary to maximize improvement of health outcomes among the population as a whole, with emphasis on those populations at highest risk for poor outcomes as a result of physical, environmental, cultural, or socioeconomic circumstances. Explore the use of nontraditional outreach strategies to reach underserved and difficult-to-reach populations.

5. Public Health Brand

Explore how specific populations, particularly those that may not be familiar or comfortable with branded organizations, perceive the public health brand [97].

Examples of Priority Research: Explore ways to measure and assess specific audiences' perception of the public health community, including perceptions of trust. Evaluate how various public health brands (e.g., CDC information, logos, and other visual representations of the agency; the brands of other federal public health agencies; and the brands of public health private health agencies, organizations, and associations) have an impact on the acceptance of information and its translation into individual and group perceptions and behavior changes. Elucidate which brands are most acceptable, explore the components of an acceptable brand, and investigate the time and circumstances required to create changes in brand perception.

6. Message Bundling

Explore strategies and practices to combine individual or similar health messages for common audiences.

Examples of Priority Research: Investigate whether combining, or "bundling," public health messages and interventions is more effective than single-issue communication for various populations and settings. ("Bundling" includes combining similar health topics that address an overarching public health issue [e.g., combining physical activity and nutrition messages for an obesity related campaign] or health issues addressing one particular audience [e.g., combining messages on the importance of vaccinations for infants with messages regarding the importance of placing infants on their backs when sleeping].) Determine whether, and under what conditions, such integrated approaches are more effective than single-issue approaches in producing changes in personal behavior, social norms, and environmental support. Evaluate possible strategies to effectively bundle individual or similar health messages for common audiences. Identify the channels, messengers, and supplementary support necessary to influence hard-to-reach populations. Determine the efficiency of bundling and the effectiveness of bundled health messages. Evaluate strategies for identifying when message bundling might be an effective health communication program strategy.

7. Emergency and Risk Communication

Explore methods of communicating health information to individuals, stakeholders, and communities during an emergency or crisis situation.

Examples of Priority Research: Evaluate key channels and strategies to enhance compliance during crises and emergencies. Assess pre-event readiness to accept and act upon specific messages through identified channels (e.g., television, radio, and the Internet) and sources (e.g., government spokespersons, subject matter experts, newscasters, and clinical practitioners). Identify strategies effectively used to encourage the general public and specified groups (e.g., health-care providers and emergency service workers) to differentiate messages from different sources and through different channels and to ultimately engage in optimal health-protection behavior. Assess post- event information management and accession efforts intended to minimize secondary health, environmental, and economic damage through promotion of social consensus and specific norms.

Evaluate key channels and strategies for translating the best available science (e.g., through formal recommendations) and for disseminating messages to target audiences to enhance compliance during crises and emergency situations.

8. Entertainment Education

Explore the use of entertainment media to promote health messages, especially for hard-to-reach audiences and for sensitive health topics.

Examples of Priority Research: Evaluate the most effective methods for using entertainment-associated health education to reach targeted audiences. Investigate how entertainment-associated health communication messages can influence knowledge, attitudes, and health-related practices. Determine which levels and types of exposure to entertainment-based health communication are associated with the adoption of positive health practices. Investigate not only the way in which health issues and concerns are portrayed, but how they are integrated into entertainment programming.

Evaluate specific theories that are most associated with developing effective entertainment-based health communication messages. Study the ways on-screen characters are portrayed as coping with health problems, and identify the methods that are most effective for modeling. Determine a) which types of television characters are considered credible sources of health information, b) which types of media children most often seek for entertainment and the opportunities for synergy among media channels, and ways youth can be most effectively targeted through entertainment-based health communication.

> We can leverage the power of the media to ensure that more children receive health and safety messages

9. Health Literacy and Clear Communication

Explore strategies and approaches for developing health literacy practices that can be incorporated within health marketing campaigns.

Examples of Priority Research: Create and improve the usability of health messages targeting non-literate, low-literate, and non-English speaking populations (as well as the public at large) to implement scientifically sound and medically accurate messages, including those that are text and non-text based. Develop tools for use in health communication interventions that allow the health literacy capacity of messages to be evaluated and measured. Design methods for testing and evaluating health literacy issues within health communication messages. Identify and develop best practices for communicating with populations that have limited health literacy.

10. Customize Health Marketing Campaigns

Explore strategies and practices for developing effective health marketing campaigns customized to specific groups, particularly those with health disparities.

Examples of Priority Research: Develop approaches and strategies to identify and segment populations with health disparities. Create, test, and evaluate customized public health messages and campaigns. Evaluate strategies for identifying the most effective communication programs for reaching various segmented audiences.

Chapter IX

PROMOTE CROSS-CUTTING HEALTH RESEARCH

The world is facing numerous health challenges associated with emerging infectious diseases, avian influenza, obesity, and natural and human-made disasters, placing increasing demands on public health agencies in the face of diminishing resources.

To meet these challenges, interdisciplinary, cross-cutting research is urgently needed to help public health professionals make better use of limited resources.

Cross-cutting research is fundamental to understanding critical infrastructure components and gaining the knowledge needed to ready the public health system for the future. This research is needed to identify and evaluate best practices and strategies to strengthen the public health workforce; cross-cutting research also will help reveal how best to use research findings to guide public policies and programs that will lead to faster, more effective ways to improve health in diverse communities. Cross-cutting research needed to elucidate ways to eliminate racial and ethnic health disparities will span across scientific disciplines and engaged partners throughout the public health community. Implementation research is especially important in determining how and why interventions do or do not work, whereas translation research will facilitate promotion and wide-spread adoption of effective interventions and sustainable community-level approaches for addressing emerging health challenges.

Cross-cutting research supports a systems approach to research that builds on existing strengths; captures interdisciplinary contributions; and promotes and enhances synergy, teamwork, and ethical integrity. Such research supports all of the Health Protection Goals developed by CDC. Ultimately, this research will help ensure that all people, especially those who experience health disparities, will achieve their optimal lifespan and experience the best possible health in every stage of life. The cross-cutting research described in this chapter spans the spectrum of priority research areas and includes:

A. Social Determinants of Health and Health Disparities
B. Physical Environment and Health
C. Health Systems and Professionals
D. Public Health Science, Policy, and Practice
E. Public Health Education and Promotion
F. Human Genomics in Public Health
G. Mental Health and Well-Being
H. Law, Policy, and Ethics

A. SOCIAL DETERMINANTS OF HEALTH AND HEALTH DISPARITIES

Although the overall health of the U.S. population has improved over the last few decades, many segments of the population still suffer from poor health (98). Developing methods to measure and compare the health status of various population groups and the relative burdens of specific diseases, injuries, and other adverse health conditions can facilitate appropriate implementation of health protection interventions. Reducing and preventing the greatest amount of total health burden across population groups is essential to improving overall health status in the United States.

Social determinants of health are those factors beyond individual behavior and genetic endowment that impact individual, community, and societal health. These factors include neighborhood conditions and resources (e.g., food supply; economic and social relationships; housing; employment; air and water quality; and the availability of transportation, education, and health care) for which distribution across populations effectively determines length and quality of life. Many of the social determinants of health exist outside of the formal health sector, even though they exert profound impacts on health status. Population and geographic distributions of the social determinants of health are not random. Structures, policies, practices, norms, and values can result in the differential allocation of resources and risks by race/ethnicity, social class, geography, sex, and other attributes of population disparity.

Achieving optimal health and quality of life for all populations, locally and globally, is contingent on collective efforts to achieve the public health goal of preventing and eliminating health disparities and ensuring that all people, especially those at greater risk of health disparities, achieve their optimal lifespan and experience the best possible quality of health in every stage of life. Health disparities (i.e., differences in the quality of health and health care that exist among population subgroups) can be observed across a wide variety of health conditions and tend to arise among populations who differ in exposure to health-impacting resources and risks. Race/ethnicity, socioeconomic status, geography, sex, age, disability status, and sexual orientation are all risk markers associated with health disparities. Further, health-care professionals have been identified as having biases toward population groups other than their own [98]. To promote behavior change among health professionals and contribute to the elimination of health disparities, research activities that focus on training to reduce these biases must be supported [99].

Research to understand the determinants of health disparities, elucidate the differences in burden of disease among subpopulations, and identify evidence-based strategies to prevent and eliminate health disparities in the 21st century must be planned, overseen, and conducted through the collaborative efforts of public health experts worldwide. By encouraging and implementing aggressive research efforts that address the risk markers for health disparities, the nation can achieve better health for all of humanity [100].

> For many of our nation's poor, good health is a luxury they cannot afford. Efforts to address the social and economic factors that influence health can ensure that all Americans have a fair chance at a healthy life.

1. Burden of Disease

Measure burden of diseases and risk factors for diseases to reduce or prevent the greatest amount of total health burden.

Examples of Priority Research: Develop and improve methods to assess current and predict future trends for health burdens, including acute and chronic illnesses and conditions, death, quality of life and well-being, and economic and social costs. Develop new methods to estimate the preventable burden of disease to ensure the appropriate allocation of resources to the most effective public health interventions. Determine methods of ascertaining and mitigating the unanticipated adverse effects of public health efforts and recommendations (e.g., increased foodborne illness resulting from public health messages to increase fruit and vegetable consumption).

2. Social Determinants of Health

Evaluate and modify social determinants of health disparities.

Examples of Priority Research: Identify and address the fundamental causes of adverse health outcomes and health disparities. Evaluate the social determinants of health by examining differences in the distributions of exposures, opportunities, and outcomes by race/ethnicity, socioeconomic status, geography, age, sex, disability, and sexual orientation.

Develop and test measures for the collection and use of standardized data to correctly identify the social determinants of health and health disparities. Develop and evaluate surveillance methods for social class markers (e.g., racism, poverty, and sex discrimination). Test and evaluate interventions that address social determinants of health disparities, including those aimed at improving high school graduation rates (e.g., after-school mentoring programs, time-management initiatives, and interventions to prevent unintended pregnancies).

Conduct studies to assess and reduce the impact of social forces (e.g., racism, sexism, and homophobia) on health. Conduct epidemiologic studies on the incidence, prevalence, and modifiable social determinants and differential risk of disease, injury, and disability in disadvantaged populations (e.g., the homeless and immigrants). Develop assessment tools that enable determination of the impact of social determinants of health and health disparities that can be used to inform new policies and programs. Assess public health surveillance systems to evaluate how they currently capture social determinants of health and how they can be improved.

Evaluate the structures, policies, practices, and norms that differentially force different populations into various contexts (e.g., the mechanisms of institutionalized racism). Evaluate the delivery of culturally competent programs in communities. Implement, track, and evaluate the dissemination of public health interventions in differentially affected populations.

3. Health Disparities Prevention and Elimination

Develop, evaluate, and promote methods and interventions that have been shown by evidence-based studies to be effective in characterizing, reducing, or eliminating modifiable health disparities.

Examples of Priority Research: Evaluate the fundamental causes of adverse health outcomes and health disparities in disadvantaged populations, and assess the need for community-level resources to remedy the primary causes of health disparities. Assess and increase the effectiveness, prevalence of use, opportunities for optimal use, strategies to increase use, and cost-effectiveness associated with interventions to prevent and control leading causes of premature death, illness, and disability in disadvantaged populations Examine health disparities among people with disabilities, including those associated with race/ethnicity, sex, and economic and employment status. Assess the use of cultural competency, language services, provider best practices, and community-based healthcare in underserved populations (e.g., the homeless and undocumented immigrants). Identify and evaluate the dissemination of successful federal, state, tribal, and local strategies that are likely to increase awareness and use of effective interventions in disadvantaged segments of the U.S. population. Identify the socioeconomic, geographic, and social barriers (e.g., stigma) that impede access to needed and quality care, and evaluate strategies aimed at improving health infrastructure to ensure equitable, population- wide access to care. Examine the role of historical urban renewal efforts in current health disparities, and determine the long-term consequences of community-level spatial reorganization on overall community health. Evaluate the role that schools and communities play in reducing or eliminating health disparities. Determine the economic and social benefits associated with reducing health disparities (e.g., assess the costs to society for continued disparities among populations with access to health care, and those who cannot access the current health system). Identify the social, physical, and mental health needs of incarcerated, and other disadvantaged persons. Evaluate efforts to capture data for under-represented groups; and implement targeted community interventions that can eliminate preventable disparities among such underrepresented and underserved groups and communities. Identify science-based interventions proven to be effective in population groups, and implement and evaluate their impact in different populations, particularly those who disproportionately experience health disparities. Assess the longterm mental health consequences of experiencing health disparities. Develop research studies that evaluate cultural competency programs and training for public health professionals, health-care providers, and students in the health-care field.

B. Physical Environment and Health

Our rapidly changing global environment continues to create new and unique public health challenges. The environments in which people live, work, learn, and play influence our health in many ways—from the air we breathe to the communities we design. Although environmental impacts (e.g., global climate change) can have far reaching health consequences, much still must be learned about how these environmental changes will influence disease processes and patterns. How we use our land and design our communities (e.g., workplaces and schools) influences rates of obesity, injury, heart disease, diabetes, and many other acute and chronic diseases and conditions. Research aimed at elucidating the relationship among the built, natural, and social-cultural environment and health allows for a comprehensive approach to solving some of public health's most challenging problems and reducing health disparities.

> Learning more about the relationship between health and the places where people live, work, learn, and play can better prepare us for the challenges of the 21st century.

1. Global Climate Change

Identify and assess the potential global and domestic health impacts posed by climate variability and change.

Examples of Priority Research: Increase the understanding and awareness of the potential human health consequences resulting from climate variability and change, with a focus on waterborne, foodborne, and vectorborne disease; morbidity and mortality from extreme weather events; air pollution and respiratory health; and complex emergency responses. Assess the impact of climate variability and change among vulnerable populations. Examine the impact of diverse U.S. climates (e.g., arctic, mountain, and desert climates) on health. Collaborate with other agencies to use Global Earth Observation System of Systems (GEOSS) (101) to improve understanding and information regarding the relationship between Earth processes and environmental health.

2. Natural and Built Environment and Health

Determine the relationships among land-use policy, the built environment, and human health and study how other critical infrastructure systems function to impede or improve public health.

Examples of Priority Research: Increase the understanding of the impact that the quality and design of our homes, schools, workplaces, communities, parks, green spaces, and transportation systems have on health throughout all stages of life, across various communities, and among persons with various types of disabilities. Examine how land use and transportation decisions can help or hinder the creation and maintenance of healthy and safe communities. Examine a variety of environmental and safety hazards, including a) substandard housing, b) indoor environmental health and safety hazards, c) poor air and water quality, d) inadequate water and sewer infrastructure; e) noise, f) inadequate pedestrian and bicycle safety, and g) poor community walkability. Examine the relative impact of each of these areas and assess the effectiveness of social policy and regulatory approaches in addressing community risks. Examine the role of public policy (e.g., zoning and environmental regulations) and social determinants of health and injury in land-use decisions and urban planning. Investigate the relationship between the built environment and mental health. Examine the impact of deforestation and loss of biodiversity on human health.

3. Physical and Sociocultural Environment and Health

Examine the relationships involved in the sociocultural environment (e.g., race, social class, and immigration) and the physical built environment, and identify health disparities.

Examples of Priority Research: Examine the relationships among decreases in housing supply and homelessness and poor mental and physical health. Conduct a comparative study to understand the impact of displacement and gentrification on mental and physical health. Examine policies and laws to understand their influence on the socially structured built environment. Examine the way segregation affects place of residence. Evaluate the effect of health-related policy on mental and physical health in communities affected by displacement and gentrification (e.g., Harlem and New Orleans). Elucidate the way health outcomes and characteristics of the built environment are affected by shrinking health resources. Investigate the relationship between concentrations of toxic waste and health status in communities of color.

C. HEALTH SYSTEMS AND PROFESSIONALS

Vital to maximizing current and future public health impact is an educated, knowledgeable workforce operating in a model public health system. A health workforce that is effective and efficient, diverse, well-educated, and committed to reaching persons at highest risk for disease, injury, and disability is key to the success of any public health effort. The health workforce must have expert knowledge and skills in a variety of areas, such as a) cultural competency, health literacy, ethics, surveillance, and strategic analysis; and b) program and policy development and evaluation. In addition to attracting and maintaining a capable health workforce, both public and private health systems must maintain performance standards, provide the best care and services to all persons in need, and be able to anticipate and respond to evolving health issues. To achieve this ideal and meet the health challenges of the 21st century, the characteristics of high-performance health systems operating at private, local, state, and federal levels must be identified.

In addition, the conventional concept of the public health system must be broadened to include diverse, capable partners and sectors. Agencies, universities, state and local health departments, and private sector practitioners must collaborate to generate new knowledge and practices that extend to each segment of the U.S. population.

> We can maximize the return on our healthcare investment by leveraging CDC's unique expertise, partnerships, and networks to improve the health system.

1. Health Systems

Study how public health systems can function to improve public health

Examples of Priority Research: Evaluate how public health systems affect public health by researching the functioning of federal, state, local, and tribal health systems. Determine how interrelationships among federal, state, and local public health agencies affect agency performance and health outcomes. Define and quantify dimensions of public health systems, including relationships among organizations. Determine how public health agency structure affects performance. Explore the relationship between health systems' performance and health outcomes. Define and characterize high-performing federal, state, tribal, and local public health agencies, including community health centers. Evaluate the factors that lead to a more effective and cost-effective public health system and maximize health outcomes per dollar invested. Evaluate the economic, cultural, social, and structural determinants of access to public health services. Identify public health gaps in service and determine ways to integrate service across programs and various levels of government.

2. Health-Care Delivery Systems

Study effective ways to work with, collect data from, and impact health-care delivery systems.

Examples of Priority Research: Evaluate how health systems affect public health by examining the functioning of private and community health-care delivery systems. Evaluate the economic, cultural, social, and structural determinants of access to medical care within the health-care delivery system. Evaluate how regional hospitals and emergency medical services can be coordinated to better respond both to incidents widely impacting community populations and to discrete mass casualty events.

3. Workforce and Career Development

Conduct research to improve public health workforce-related recruitment, retention, and training.

Examples of Priority Research: Define and assess the knowledge, attitudes, and skill competencies needed by the public health workforce, and identify information gaps resources, and other areas in need of improvement. Identify strategies to create and sustain career development and entry-level supply lines to ensure the existence of a workforce pool capable of meeting the increased demands on public health. Identify best practices for workforce recruitment, retention, and training. Identify and apply new methods to describe current public health workforce needs and forecast future needs. Assess the impact of trained public health professionals on specific outcomes (e.g., improved health of people and improved public health practice or capacity). Identify best practices for workforce development. Determine the best methods of facilitating collaboration among academia and other groups (e.g., community-based organizations) and integrating efforts with public health practice. Determine the best methods for identifying, updating, and validating competencies necessary for an effective, efficient health workforce. Determine the organizational variables that support development and application of skill competencies. Determine best indicators for workforce performance. Monitor workforce trends, including size, distribution, qualifications, and tenure. Evaluate the role of labor market forces on recruitment, retention, wage, salary, benefits, and personnel-system characteristics.

D. PUBLIC HEALTH SCIENCE, POLICY, AND PRACTICE

CDC and its partners in public health aim to improve the health status of all persons living in the United States, reduce health disparities among subgroups of Americans, and work with nations and international agencies to extend the benefits of collaborative research and effective programs to populations around the world. Public health research and community-based participatory research (CBPR) help discover the biological, social, anthropological, cultural, and behavioral determinants of health; uncover how genetic, environmental, and other forces interact to produce biological and social vulnerability; describe health disparities; and reveal which interventions are effective in particular settings or populations. However, the key to sustained, organized efforts to improve the health of specific populations is having knowledge beyond that of basic intervention effectiveness.

More research is needed to discover how best to identify the burden of disease, identify best practices for promoting health and improving quality of life, evaluate and measure the impact of interventions, disseminate effective interventions beyond initial research or test settings, remove barriers to the adoption of research findings into practice, and adapt interventions that have been shown to work in specific places to render them effective in different populations and in diverse settings. One approach to achieving effective interventions is CBPR. CBPR is a collaborative research process between researchers and community representatives that engages community members, derives understanding, serves as the basis for interventions generated from local knowledge of health problems, and invests community members in all stages of research [102].

A new and emerging area of interest is complementary and alternative medicine (CAM). An estimated one third of the U.S. population uses some type of CAM (103,104). Research to better illuminate the efficacy and applicability of complementary and alternative therapies also should be conducted.

1. Intervention and Translational Research

Develop and evaluate strategies to translate, disseminate, and sustain science-based interventions, and identify best practices to promote health and quality of life by preventing and controlling disease, injury, and disability.

Examples of Priority Research: Develop and test models for the translation, dissemination, and institutionalization of effective programs and policies to promote health and prevent disease, injury, and disability in diverse populations. Identify methods that can be used to accelerate the adoption of science-based programs, policies, laws, and behaviors at the individual, family, community, business, public health practice, organizational, and social policy levels. Investigate means of ensuring the culturally competent delivery of effective interventions. Develop effective strategies to improve knowledge transfer from the public to the private sector.

2. Social, Anthropological, and Behavioral Sciences in Public Health

Develop and apply social and behavioral theories and methods to assess and improve public health at individual, family, community, institutional, tribal, regional, state, federal, and international levels.

Examples of Priority Research: Develop and evaluate methods for describing and analyzing behavioral, cultural, and social determinants of health (e.g., poverty, housing, racism, education, and social capital). Develop and evaluate indicators of baseline community health status. Establish methods and standards that employ baseline data to evaluate the effectiveness of health interventions in improving people's health. Evaluate quantitative, observational, and qualitative research methods (e.g., rapid ethnography) for use in the surveillance of social, anthropological, and behavioral factors associated with adverse health outcomes. Develop and evaluate culturally competent social, anthropological, and behavioral science interventions to prevent disease, injury, and disability and to promote health and reduce disparities. Identify individual, community, and institutional factors that result in preventive health-care practices. Investigate methods used to change health behaviors (e.g., those that result in optimal acceptance and utilization), and translate those methods into practical, sustainable programs.

3. Health and Well-Being Across Diverse Community Settings

Develop and implement strategies that enable individuals, families, schools, health systems, and employers in all communities to promote and sustain health for all of their residents.

Examples of Priority Research: Identify and measure the impact of social, economic, legal, environmental, and behavioral factors (e.g., smoking and other types of substance misuse and risky sexual behavior) on health in community settings. Evaluate the effectiveness of laws, policies, and incentives (e.g., water fluoridation, bike paths, immunizations, hand washing, and clean air) designed to improve health. Identify, implement, and evaluate effective health promoting approaches targeted to community settings (e.g., encouraging active lifestyles, healthy eating, and tobacco cessation), and implement approaches designed to prevent or delay the onset of related chronic and infectious diseases, injuries, and disabilities. Conduct needs assessment of resources and regional planning. Conduct economic analyses of health promotion, disease control, and injury prevention activities to estimate the economic burden of disease. Identify and evaluate interventions that produce the greatest return on investment, and provide a rationale for prioritizing resources at the community level.

4. Economics and Public Health

Develop and apply economic theories and methods to examine the supply and demand for public health services; estimate the monetary and social costs of diseases and injuries; determine the return on investment for public health interventions; and improve the delivery of health-care prevention, treatment, and rehabilitation services.

Examples of Priority Research: Develop methods to address gaps in information for various economic measures of health burden and well-being.

Study the effectiveness of interventions to inform resource allocation, and determine the economic return on investment for various public health interventions. Develop models to illustrate how public health resources and funding can most efficiently be used to optimize the public's health and how federal, state, business, industry, and local jurisdictions can best share the costs of common health problems (e.g., infectious disease among immigrants). Conduct applied economic studies of factors (e.g., delayed gratification, risk perception of behaviors, preventive health interventions, social and environmental factors, and risk-taking preferences) that can affect individual and organizational choices.

5. Community-Based Participatory Research (CBPR)

Design, conduct, and interpret CBPR research to increase the relevance, acceptability, and usefulness of evidence-based scientific findings in improving the health of communities.

Examples of Priority Research: Evaluate the influence of CBPR on improvements in community health status and reduction of health disparities, especially in disadvantaged, minority, and hard-to-reach populations. Identify barriers to and opportunities for successful engagement of communities in research. Study and adapt evidence-based practices for use in diverse populations. Conduct CBPR to understand and evaluate interventions at the community level, especially those that focus on policy and environmental changes aimed at reducing the burden of and risk factors for chronic conditions and increasing the dissemination of effective health-promotion interventions. Identify ways to make interventions sustainable within communities. Evaluate comprehensive community interventions using, for example, an approach similar to that employed by the *Guide to Community Preventive Services* [13], and identify research gaps. Enable community health status to be assessed in new and comprehensive ways by evaluating the effectiveness of strategies to increase community involvement (e.g., community members [particularly those affected by illness and disease], partners, and stakeholders, including those with regional and urban planning expertise) in the planning, translation, and execution of health research and interventions. Evaluate the impact of dissemination of research findings and knowledge to the community.

6. Public Health Evaluation Research

Improve evaluation methods, and conduct evaluations on the public health impact of interventions, programs, and policies.

Examples of Priority Research: Evaluate the processes, context, impact, and outcomes associated with federal, state, tribal, and local health intervention efforts. Evaluate the relationship between adverse health outcomes and a) health insurance coverage and b) access to medical care, and compare the relative benefit of expanded coverage to improve the public's health. Evaluate how public health assessment, surveillance, intervention, and service delivery programs can be optimally structured to deliver the highest quality and widest array of proven effective activities to meet community health needs. Conduct health impact assessments (105) for policies and programs not explicitly designed to address health issues (e.g., policies for housing, transportation, education, labor, early childhood development, and reduction of discrimination), and develop methods for evaluating these programs and policies. Further investigate the impact of evaluation research.

7. Use and Allocation of Pharmaceuticals and Other Medical Interventions

Investigate the way allocation and use of pharmaceuticals and other treatment and intervention options affect public health and the health-care industry.

Examples of Priority Research: Optimize the availability of drug agents, and promote the availability and appropriate use of pharmaceuticals and other medical treatments and interventions to improve public health. Examine the effect of supply and demand, and investigate the causes of market failure.

8. Complementary and Alternative Medicine (CAM)

Assess the use of CAM modalities and their role in effective prevention and treatment of adverse health conditions.

Examples of Priority Research: Evaluate the prevalence and demographics of use of CAM modalities for the treatment of various health conditions. Working with partners, evaluate CAM regimens to determine their safety and efficacy in treating various infectious and non-infectious conditions. Develop and test scientifically-based guidelines for practice- based assessments of CAM regimens. Evaluate effective strategies for educating the public about CAM. Conduct translation and dissemination research on CAM interventions and treatments that have been shown to be effective in the general population, and examine the use of CAM modalities among individuals seeking care from traditional and non-traditional health-care providers.

> We can improve the health of our nation by enabling more Americans to make healthier, safer choices

E. PUBLIC HEALTH EDUCATION AND PROMOTION

Public health education and health promotion play a critical and broad role in preventing illness, injury, disability, and premature mortality. These activities educate individuals, communities, organizations, and institutions about the principles, concepts, and interventions associated with public health. However, the potential of health education to influence healthy choices and changes in behavior is often under appreciated. Research on health education and promotion is needed to support evidence-based approaches to promote the health and well-being of the general U.S. and global populations, as well as specific populations at risk. Research in health education and promotion can also lead to reductions in healthcare and other societal costs. Health education and promotion represent a combination of approaches, methods, and strategies drawn from diverse disciplines (including behavioral, social, and health sciences). Health education applies theory and scientific evidence and incorporates the values of social justice and individual and community's perceptions of health; it focuses on the knowledge, attitudes, beliefs, values, and interests of people, communities, and organizations and builds on these to impact social and behavioral change at multiple levels (e.g., individual, family, community, and organizational). To adequately address and reduce health disparities in diverse populations, health education requires collaboration in diverse cultural, community, and organizational settings.

Advocacy, policy change, and organizational change are central activities of public health education and health promotion. Health education of both public-

and private-sector health-care providers serves as a fundamental means to achieving public health objectives and improving the success of public health interventions.

1. Public Health Education Research

Increase understanding regarding the way health education strategies and programs impact health and health behaviors.

Examples of Priority Research: Identify methods to define and measure intermediate and long-term outcomes of health education programs, policies, and other activities. Assess the impact of health education that aims to change social norms, policy, and environments on health behaviors and health disparities. Examine the relationship between environmental changes and health education and policy, and explore the way this relationship produces behavioral change. Identify and evaluate strategies and venues that can best be used to deliver health-education interventions (e.g., community-, group-, individual-, and interactive session-based education and electronic/web-based education). Evaluate health education materials and messages for cultural competency and effectiveness within specific populations across the lifespan. Assess knowledge and awareness of health-related information, policies, and regulations among persons in various settings (e.g., workplaces, schools, and health-care facilities). Increase understanding of the public health education needs of target audiences, particularly persons who are disproportionately at risk for disease, injury, and disability. Assess and understand how health and science literacy among the general public, health-care providers, legislators, and others impacts the health of various populations. Determine best practices to improve health literacy. Assess the impact of health education on individual, organizational, and community health behavior during routine and emergency situations. Evaluate use of daily lifestyle profiles (i.e., descriptions of people's daily activities and behaviors) as indicators of health risk level for certain groups, and determine the potential of these profiles to appropriately identify interventions that are likely to be accepted and adopted within specific populations. Develop and evaluate culturally competent methods to effectively communicate risk to diverse communities to inform individual, community, and legislative health decisions, and examine the effectiveness of communication tools (e.g., back-translation practices). Develop theory- and evidence-based curricula to be used in primary, secondary, and higher education programs and in post-graduate and community education. Carefully test, refine, and disseminate health-education theories to inform innovative program development and evaluation. Evaluate the economic impact of health education policies and practice Determine methods of effectively countering anti-public health initiatives (e.g., those that encourage tobacco use).

> Discoveries in human genetics hold the promise of better health for all. Only through joint efforts of genomics research and public health can we harness the potential of the human genome to prevent illness and save lives.

F. HUMAN GENOMICS IN PUBLIC HEALTH

Human genomics is the study of human genes and how they interact with each other and the environment. The use of human genomics in public health research will help determine why some people become ill (both physically and mentally) from certain infections, environmental exposures, and behaviors while others do not. Better integration of information on inherited, environmental, and behavioral risks will help tailor intensive health promotion and disease prevention strategies to groups most in need, which ultimately can reduce health disparities (106). Public health scientists are working towards translating genomics research in a way that benefits individuals, families, and communities nationwide while examining the relevant ethical, legal, and social implications of this research from a public health perspective. Only through broad-based research in diverse communities can the full benefits of human genomics be realized.

1. Genomics Bridge Between Preventive Medicine and Public Health Research

Integrate population-level data on genotypes, environmental and social risk factors, family history, and specific disease and mental health outcomes to identify high-risk communities, families, and individuals for intensive intervention.

Examples of Priority Research: Evaluate the validity and utility of family history and genetic testing (including newborn, family-based, and population screening). Evaluate trials of community- and family-based interventions that address both inherited and shared environmental and behavioral factors.

2. Human Genomics, Public Health Surveillance Systems, and Research Studies

Collect genomic and environmental information derived from surveillance systems and epidemiologic studies to examine gene-environment interactions and identify environmental targets for intervention.

Examples of Priority Research: Demonstrate and evaluate the appropriate integration of family history and genomic markers (including genotype, gene expression, and proteomic data) into surveillance systems (e.g., cancer registries), and evaluate the benefits of collecting genomic information in epidemiologic studies that focus on all aspects of health (i.e., physical and mental).

3. Models to Incorporate Education and Community Engagement into Population-Based Genomics Research

Develop and evaluate methods for engaging communities in assessing the risks, benefits, and tradeoffs associated with research participation.

Examples of Priority Research: Develop and evaluate methods for identifying and measuring potential risks and benefits of genomic research participation for individuals, families, and communities. Examine strategies for communicating research objectives, benefits and risks, and results. Assess tradeoffs between individual and social concerns (e.g., privacy and information sharing). Examine ways to protect the security of genomic data and other host factors.

4. The Genomic Evidence Base in Public Health Practice

Develop sound methods and practical tools for collecting, analyzing, and synthesizing genomic information in populations and for evaluating the public health, ethical, and legal impact of these types of data.

Examples of Priority Research: Identify and develop effective methods and tools for collecting, analyzing, and synthesizing information on genetic variation, gene-disease associations, gene-environment interactions, and genetic tests. Identify and develop studies to examine ethical challenges, legal issues, and social implications of genomics research from a public health perspective.

5. Human-Pathogen Genomics

Investigate genetic and immunologic features of human-pathogen interactions to understand and prevent infectious causes of acute and chronic diseases.

Examples of Priority Research: Investigate human and pathogenic genetic and immunologic factors associated with a) susceptibility to disease transmission and natural history, b) amenability to prophylaxis, c) treatment response, d) priority groups for intervention, e) adverse long-term sequelae, f) drug resistance, and g) side effects of drugs and vaccines.

G. MENTAL HEALTH AND WELL-BEING

Mental health plays a primary role in ensuring the overall health and well-being of all persons. Poor mental health (along with the often co-occurring substance-use disorders) significantly contributes to increased rates of injuries, chronic and infectious disease, family violence, underemployment, and litigation. The Surgeon General's Report on Mental Health (107) indicated that although much is known about how to treat mental illness, additional knowledge is needed regarding ways to prevent mental illness and promote health. Various approaches can be used, including the evaluation of neuropsychological effects and cognitive function.

Mental health and substance-abuse prevention and treatment are vital components in the elimination of health disparities. Although poor mental health and substance abuse are concerns for people in all stages of life, they disproportionately affect certain populations. For instance, persons who live in inner-city and rural settings, the homeless, and other persons who receive basic medical care in emergency departments often do not receive treatment for mental health and substance abuse (107). Emergency department settings are typically overcrowded and thus limit patient access to basic care and appropriate referral services. Further, many primary-care practitioners do not screen for substance-use disorders or mental health problems. Stigma associated with alcohol and other drug problems, as well as mental health problems, can result in lack of identification and treatment of these problems; therefore, efforts to better promote assessment and interventions for these types of disease are critical. Research that elucidates the best methods of preventing alcohol and other drug problems will help protect the health and safety of all persons within a community. Models for incorporating effective mental health and substance-use control strategies into community-based programs and medical-care settings can save lives; reduce stress; and prevent excess disability, underemployment, and litigation.

1. Mental Health

Determine the influence of mental health on overall health and well-being, and identify and evaluate best practices for prevention, assessment, and treatment of mental illness.

Examples of Priority Research: Identify the most effective individual-, family-, and community-level interventions for preventing, diagnosing, and treating mental illness at all stages of life. Examine the behavioral determinants of mental health and neuropsychology. Conduct epidemiologic studies using a variety of approaches (including the evaluation of neuropsychological effects and cognitive function) to increase understanding of mental health in the U.S. population and to evaluate available mental health services. Evaluate the social and economic costs of mental illness, depression, violence, suicide, and post-traumatic stress on individuals, families, and communities. Evaluate strategies aimed at motivating persons to seek treatment for mental illness, reducing barriers to access for such treatment, and improving the ability of primary-care providers to access and treat mental illness. Evaluate mental health assessment and treatment strategies, including the integration of mental health treatment into other medical-care settings (e.g., emergency departments and primary health-care settings). Examine the impact of co-morbid conditions on mental health outcomes. Examine the practices of physicians who prescribe psychotropic medications (e.g., antidepressants, antipsychotics, and stimulants), and elucidate the long-term effects of these medications on specific populations, particularly children. Identify strategies and interventions that help victims and first responders cope with the mental health consequences of traumatic events (e.g., war, natural disasters, rape, childhood neglect, and childhood physical or sexual abuse). Evaluate the need for and existing skills of persons working in public service (e.g., law enforcement officers, teachers, and emergency responders) who may interface with people who have mental illness. Investigate the relationship among mental, spiritual, and physical health using large-scale population data (e.g., National Health Interview Survey [108]).

2. Substance Abuse Treatment and Prevention

Determine the influence and impact of substance abuse on overall health and well-being for individuals and communities, and identify and evaluate best practices for prevention, assessment, and treatment of alcohol and other types of substance abuse.

Examples of Priority Research: Evaluate the influence of alcohol and other drug problems on illness, injury, and mortality. Determine the societal and economic costs of and the burden of disease resulting from alcohol and other drug problems, including loss of productivity, decreased family and social functioning, crime, violence, and other associated health problems. Examine the relationship between substance abuse and mental health. Evaluate strategies to prevent or reduce the negative consequences of alcohol misuse on health and safety, and investigate the prevention of underage drinking. Evaluate long- and short-term strategies aimed at preventing substance abuse, including harm-reduction strategies and those that are policy and school based. Evaluate the impact of substance-abuse assessment and treatment strategies, including the integration of substance-abuse treatment into medical-care settings (e.g., emergency departments, mental health clinics, and primary-care settings). Evaluate the characteristics of successful prevention and treatment programs (e.g., type of program, process used, implementation, funding level, and program length). Determine the effectiveness and cost-effectiveness of substance-abuse prevention strategies (including those that focus on illicit drug use, misuse of prescription drugs, and smoking). Investigate the epidemiology of drug use among young adults, and develop model prevention strategies for use in this age group. Increase the knowledge-base regarding the gender- and age-based differences associated with substance-abuse prevention and treatment programs.

H. LAW, POLICY, AND ETHICS

Law, policy, and ethics are integral to the foundation of public health research and to the practice of public health in every domain. Law has been a critical tool throughout the history of organized public health, as it has been used to authorize and shape public health programs and services. Most public health interventions rely heavily on legal elements and tools; many public health interventions take the form of law (e.g., bicycle helmet laws and niacin food supplementation requirements). Policies (both public and private sector) play an important complementary role. Research on law and policy as public health tools spans a wide spectrum of domains and priorities. Specifically, research on the role and impact that law and policy have on public health can determine how these tools can best be used to promote health and eliminate health disparities. Ethical conduct also is critical to public health efforts; ethics ensure that research and interventions achieve the highest standards of credibility, equity, and effectiveness. Research on public health ethics will improve public health research and can be used to inform efforts to translate findings into effective practice. Additional research on law, policy, and ethics is needed to strengthen CDC's public health practice, inform policy- related decisions, and ensure the conduct of sound, meaningful research.

1. Public Health Policy and Law

Determine the optimal role for law and policy as public health tools, and improve the translation of research findings at the community, state, national, and international practice levels.

Examples of Priority Research: Evaluate the ways laws, regulations, and policies contribute to the prevention of morbidity and mortality and to the reduction of health disparities. Evaluate the effectiveness of laws and policies as public health interventions, and evaluate the effectiveness and interpretations of local laws and policies concerning public health interventions across jurisdictions. Identify factors that contribute to their effectiveness, and identify their interaction with legal rights and principles. Develop the methodological basis for research on law and policy as public health tools. Identify methods to effectively translate scientific findings into information for use by public health providers and policymakers. Examine the role of public health as a foundation for effective law. Examine the impact of international health regulations on public health infrastructure. Conduct documentation studies designed to illustrate how various legal approaches impact public health policy and practice.

2. Ethics in Public Health

Determine the best methods of improving the development and promoting the adoption of ethical frameworks and practices in the conduct of public health research and programs.

Examples of Priority Research: Develop, implement, and evaluate educational tools and trainings for incorporating ethical principles into public health research, policy, and practice. Develop and evaluate the ethical competency level and training needs for public health professionals. Evaluate the national infrastructure and resources needed for ethics education and training (e.g., ethics courses in schools of public health). Evaluate the health benefits of promoting and protecting human rights in public health research, policy, and practice. Analyze the human-rights issues associated with health and health care. Evaluate methods to enhance the protection of privacy and confidentiality in public health programs and research, with emphasis on balancing privacy and data collection with public health goals. Develop tools for assessing and minimizing the potential impact of public health programs and research on human rights. Evaluate the ethics of conducting research in community settings, and develop methods to translate those findings to protect human rights. Identify and examine ethical considerations (e.g., the distribution of medication) during preparedness-response activities and large-scale public health emergencies. Develop indicators and tools for measuring the potential health impact of ethical and human rights violations in public health research, policy, and practice. Investigate the ethical considerations associated with health-promotion activities, including advertising.

REFERENCES

[1]   U.S. Department of Health and Human Services. Healthy people 2010: understanding and improving health [electronic version]. 2nd ed. Washington, DC: U.S. Government Printing Office, 2000. Available from URL: http://www.healthypeople.gov/Document/tableofcontents.htm#under.

[2]   Scheld WM, Murray BE, Hughes JM, eds. Emerging infections. Washington, D.C.: American Society for Microbiology Press, 2004.

[3]   Institute of Medicine. Emerging infections: microbial threats to health in the United States. Washington, DC: National Academy Press, 1992.

[4]   CDC. Bioterrorism agents/diseases [online]. Available from URL: http://www.bt.cdc.gov/agent/agentlistcategory.asp.

[5]   Thompson WW, Shay DK, Weintraub E, et al. Mortality associated with influenza and respiratory syncytial virus in the United States. JAMA 2003;289:179--86.

[6]   The White House. National strategy for pandemic influenza [online]. 2005. Available from URL: http://www.whitehouse.gov/homeland/pandemic-influenza.html.

[7]   (ENDNOTE #2) Weber JT, Courvalin P. An emptying quiver: antimicrobial drugs and resistance. Emerg Infect Dis 2005;1 1(6) :791-3.

[8]   (ENDNOTE #3) Nosowsky R, Giordano TJ. The Health Insurance Portability and Accountability Act of 1996 (HIPAA) privacy rule: implications for clinical research. Annu Rev Med 2006;57:575-90.

[9]   Strom BL, ed. Pharmacoepidemiology. 4th ed. Hoboken, NJ: John Wiley & Sons, 2005.

[10] Hartzema AG, Porta M, Tilson HH, eds. Pharmacoepidemiology: an introduction. 3rd ed. Cincinnati, OH: Harvey Whitney Books Company, 1998.

[11] (ENDNOTE #4) CDC. Achievements in public health, 1900–1999: control
      of infectious diseases. MMWR 1999;48:621-29.
[12] CDC. Achievements in public health, 1900-1999: impact of vaccines
      universally recommended for children – United States, 1990-1998. MMWR
      1999 ;48(12);243-8.
[13] CDC. Guide to community preventive services [online]. 2005. Available
      from URL: http://www. thecommunityguide.org.
[14] Wallace RB, ed. Public health and preventive medicine. 14th ed. Stamford,
      CT: Appleton & Lange, 1998.
[15] (ENDNOTE #1) Armstrong GL, Conn LA, Pinner RW. Trends in infectious
      disease mortality in the United States during the 20th century. JAMA
      1999;281:61-6.
[16] WHO. WHO Report on infectious diseases: removing obstacles to healthy
      development, 1999. Geneva: WHO, 1999;68. WHO/CDS/99.1. Order No.
      19300156.
[17] Knobler SL, O'Connor S, Lemon SM, Najafi M, eds. The infectious
      etiology of chronic diseases. In: Institute of Medicine. Forum on microbial
      threats linking infectious agents and chronic diseases [workshop summary
      and assessment]. Washington, D.C.: National Academy of Sciences, 2004.
[18] Gabbe SG. Obstetrics: normal and problem pregnancies. New York, NY:
      Churchill Livingstone, 2002.
[19] Safe Motherhood. Maternal mortality [online]. 2002. Available from URL:
      http://www.safemotherhood. org/facts_and_figures/maternal_mortality.htm.
[20] Bryce J, Boschi-Pinto C, Shibuya K, Black RE, the WHO Child Health
      Epidemiology Reference Group. WHO estimates of the causes of death in
      children. Lancet 2005;365: 1147–52.
[21] Executive Office of the President. The federal response to Hurricane
      Katrina: lessons learned [online]. 2006. Available from URL:
      http://www.whitehouse.gov/reports/katrina-lessons-learned/.
[22] Olson D, Leitheiser A, Atchison C, Larson S, Homzik C. Public health and
      terrorism preparedness: cross-border issues. Public Health Rep 2005;
      120:75–83.
[23] Marmagas SW, King LR, Chuk MG. Public health's response to a changed
      world: September 11, biological terrorism, and the development of an
      environmental health tracking network. Am J Public Health 2003; 93(8):
      1226–30.
[24] United States Government Accountability Office. Homeland Security DHS'
      efforts to enhance first responders' all-hazards capabilities continue to
      evolve: report to the Chairman and ranking democratic member [online].

Washington, DC: Subcommittee on Economic Development, Public Buildings and Emergency Management, Committee on Transportation and Infrastructure, House of Representatives, 2005. Available from URL: http://www.gao.gov/new.items/d05652.pdf.

[25] Barnett DJ, Balicer RD, Blodgett D, Fews AL, Parker C, Links JM. The application of the Haddon Matrix to public health readiness and response planning. Environ Health Perspect 2005; 113:561–6.

[26] Miner K, Redmon P, Toomey K. A wake-up call for public health. Public Health Rep 2005; 120:1–2.

[27] Fraser MR, Brown DL. Bioterrorism preparedness and local public health agencies: building response capacity. Public Health Rep 2000; 115(4):326–30.

[28] Blanchard JC, Haywood Y, Bradley S, Tanielian TL, Stoto M, Lurie N. In their own words: lessons learned from those exposed to anthrax. Am J Public Health 2005; 95(3):489–95.

[29] Boyle CA, Decoufle P, Yeargin-Allsopp M. Prevalence and health impact of developmental disabilities in U.S. children. Pediatrics 1994; 93(3):399–403.

[30] Anderson G, Knickman JR. Changing the chronic care system to meet people's needs. Health Affairs 2001; 20(6): 146–60.

[31] Alliance for Health Reform. America's most ignored health problem: caring for the chronically ill [online]. 2001. Available from URL: http://208.58.24.194/pub/pdf/5_01AHRv8.pdf.

[32] Mokdad AH, Marks JS, Stroup DF, Gerberding JL. Actual causes of death in the United States, 2000. JAMA 2004; 291:1238-45. [Correction published in Mokdad AH, Marks JS, Stroup DF, Gerberding JL. Actual causes of death in the United States, 2000—correction. JAMA 2005;293:293–8.]

[33] Correa-Villasenor A, Cragan J, Kucik J, O'Leary L, Siffel C, Williams L. The Metropolitan Atlanta Congenital Defects Program: 35 years of birth defects surveillance at the Centers for Disease Control and Prevention. Clinical and Molecular Teratology 2003; 67:617–24.

[34] Shi L, Stevens GD, Wulu JT Jr, Politzer RM, Xu J. America's health centers: reducing racial and ethnic disparities in perinatal care and birth outcomes. Health Services Research 2004; 39 (6 Pt 1):1881–901.

[35] Newacheck PW, Taylor WR. Childhood chronic illness: prevalence, severity, and impact. Am J Public Health 1992; 82(3):364–71.

[36] Newacheck PW, McManus MA, Fox HB. Prevalence and impact of chronic illness among adolescents. American Journal of Diseases in Children 1991; 145(12):1367–73.

[37] National Center for Health Statistics. Health, United States, 2004 [electronic version]. Hyattsville, Maryland: Department of Health and Human Services, National Center for Health Statistics, 2004. Available from URL: http://www.cdc.gov/nchs/hus.htm.

[38] Centers for Disease Control and Prevention [online]. Health information for older adults. 2005. Available from URL: http://www.cdc.gov/aging /health_issues.htm#5 .

[39] Boyle CA, Yeargin-Allsopp M, Doernberg NS, Holmgreen P, Murphy CC, Schendel DE. Prevalence of selected developmental disabilities in children 3–10 years of age: the Metropolitan Atlanta Developmental Disabilities Surveillance Program, 1991. MMWR 1996; 45(No. SS-02):1–14.

[40] Chirikos TN. Aggregate economic losses from disability in the United States: a preliminary assay. Milbank quarterly 1989; 67(2):59–91.

[41] Brandt EN, Pope AM. Enabling America: assessing the role of rehabilitation science and engineering. Washington, DC: Institute of Medicine, National Academy Press, 1997.

[42] Kraus L, Stoddard S, Gilmartin D. Chartbook on disability in the United States, 1996: an InfoUse report. Washington, DC: U.S. National Institute on Disability and Rehabilitation Research, 1996.

[43] U.S. Department of Health and Human Services. The Surgeon General's call to action to improve the health and wellness of persons with disabilities [electronic version]. Washington, DC: US Department of Health and Human Services, Office of the Surgeon General, 2005. Available from URL: http://www. surgeongeneral.gov/library/disabilities/calltoaction /calltoaction.pdf.

[44] Web-based Injury Statistics Query and Reporting System (WISQARS) [online]. 2006. Available from URL: http://www.cdc.gov/ncipc/wisqars.

[45] Doll L, Mercy J, Bonzo S, Sleet D. Handbook of injury and violence prevention. New York: Springer, 2006.

[46] Weeks JL, Levy BS, Wagner GR, eds. Preventing occupational disease and injury. 2nd ed. Washington, DC: American Public Health Association, 2005.

[47] Melse JM, de Hollander AEM. Environment and health within the OECD region: lost health, lost money. Bilthoven, the Netherlands: National Institute of Public Health and the Environment (RIVM), 2001. Report no. 402101 001.

[48] Work-related injury statistics query system [online]. 2004. Available from URL: http://www2a.cdc. gov/risqs.

[49] Bureau of Labor Statistics. Workplace injuries and illnesses in 2003 [online]. 2004. Available from URL: http://www.bls.gov/news.release/pdf/osh.pdf.

[50] Bureau of Labor Statistics. National census of fatal occupational injuries in 2004 [online]. 2005. Available from URL: http://www.bls.gov/news.release /pdf/cfoi.pdf.

[51] Steenland K, Burnett C, Lalich N, Ward E, Hurrel J. Dying for work: the magnitude of U.S. mortality from selected causes of death associated with occupation. Am J Ind Med 2003; 43: 461–82.

[52] National Academy of Social Insurance. Workers' compensation: benefits, coverages, and costs, 2002 [online]. 2004. Available from URL: http://www.nasi.org/usr_doc/workers_comp_2002.pdf.

[53] Leigh JP, Markowitz SB, Fahs M, Landrigan PJ. Costs of occupational injuries and illnesses. Ann Arbor, MI: University of Michigan Press, 2000.

[54] Krug EG, Dahlberg LL, Mercy JA, Zwi AB, Lozano R. World report on violence and health. Geneva, Switzerland: World Health Organization, 2002.

[55] Campbell JC, Jones AS, Dienemann J, Kub J, Schollenberger J, O'Campo P, Gielen AC, Wynne C. Intimate partner violence and physical health consequences. Arch Intern Med 2002; 162(10): 1157–63.

[56] Coker AL, Davis KE, Arias I, Desai S, Sanderson M, Brandt HM, Smith PH. Physical and mental health effects of intimate partner violence for men and women. Am J Prev Med 2002; 23(4):260–8.

[57] Felitti V, Anda R, Nordenberg D, Williamson D, Spitz A, Edwards V, Koss, MP, Marks, JS. Relationship of childhood abuse and household dysfunction to many of the leading causes of death in adults. Am J Prev Med 1998; 14(4):245–58.

[58] Segui-Gomez M, MacKenzie EJ. Measuring the public health impact of injuries. Epidemiol Rev 2003; 25:3–19.

[59] CDC. CDC injury research agenda. Atlanta, GA: Centers for Disease Control and Prevention, National Center for Injury Prevention and Control, 2002.

[60] Stevens JA, Olson S. Reducing falls and resulting hip fractures among older women. MMWR Recomm Rep 2000 Mar 31; 49(RR-2):3-12.

[61] Anderson RN, Smith BL. Deaths: leading causes for 2002. National Vital Statistics Report, 2005; 53(17):1–92.

[62] CDC. CDC acute injury care research agenda: guiding research for the future. Atlanta, Georgia: Centers for Disease Prevention and Control, National Center for Injury Prevention and Control, 2005.

[63] Finkelstein E, Corso P, Miller T. The incidence and economic burden of injuries in the United States. New York: Oxford University Press, 2006.

[64] Sleet DA, Liller KD, White DD, Hopkins K. Injuries, injury prevention, and public health. American Journal of Health Behavior 2004; 28(Suppl 1):S6–S12.

[65] US Department of State. The President's plan for emergency AIDS relief: five-year global HIV/AIDS strategy [online]. 2004. Available from URL: http://www.state.gov/s/gac/rl/or/c11652.htm.

[66] United Nations. Millennium declaration: General Assembly Resolution 55/2 [online]. 2000. Available from URL: http://www.un.org/millennium/declaration/ares552e.htm.

[67] UNAIDS. 2006 Report on the global AIDS epidemic [online]. 2006. Available from URL: http://data.unaids.org/pub/GlobalReport/2006/2006_GR-ExecutiveSummary_en.pdf.

[68] Corbett EL, Watt CJ, Walder N, et al. The growing burden of tuberculosis: global trends and interactions with the HIV Epidemic. Arch Intern Med 2003; 163,1009–21.

[69] World Health Organization. Sexually transmitted infections fact sheet [online]. 2004. Available from URL: http://www.who.int/reproductive-health/rtis/docs/sti_factsheet_2004.pdf.

[70] World Health Organization. The world health report 2005 [online]. 2005. Available from URL: http://www.who.int/whr/2005/en/.

[71] Fauci AS, Touchette NA, Folkers GK. Emerging infectious diseases: a 10-year perspective from the National Institute of Allergy and Infectious Diseases [electronic version]. Emerging Infectious Diseases 2005. Available from URL: http://www.cdc.gov/ncidod/EID/vol11no04/04-1167.htm.

[72] World Health Organization. Chronic disease information sheets. 2003 [online]. Available from: http://www.who.int/dietphysicalactivity/publications/facts/en/.

[73] Peden M, McGee K, Krug E, eds. Injury: a leading cause of the global burden of disease, 2000 [electronic version]. Geneva: World Health Organization, 2002. Available from URL: http://whqlibdoc.who.int/publications/2002/924 1562323 .pdf.

[74] Peden M, Scurfield R, Sleet D, et al., eds. The world report on road traffic injury prevention [electronic version]. Geneva: World Health Organization,

2004. Available from URL: http://www.who.int/worldhealth-day/2004/infomaterials/world_report/en/.

[75] World Health Organization. Drowning fact sheet. 2005 [online]. Available from URL: http://www.who. int/violence_injury_prevention/ publications/other_injury/en/drowning_factsheet.pdf.

[76] Sasser S, Varghese M, Kellermann A, Lormand JD. Prehospital trauma care systems. Geneva: World Health Organization, 2005.

[77] World Health Organization and International Labour Office. Number of work related accidents and illnesses continues to increase [online]. 2005. Available from URL: http://www.who.int/mediacentre/news /releases/2005/pr18/en/.

[78] United Nations. 2004 Humanitarian appeals [online]. 2004. Available from URL: http://www.un.org/ depts/ocha/cap/appeals.html.

[79] International Labour Organization. Child labour statistics [online]. 2005. Available from URL: http://www.ilo.org/public/english/standards /ipec/simpoc/index.htm.

[80] Standing Against Global Exploitation (SAGE) Project. Key statistics [online]. Available from URL: http://www.sageprojectinc.org/html /info_statistics.htm.

[81] UNICEF. Child protection from violence, exploitation and abuse. [online]. Available from URL: http://www.unicef.org/protection/index_orphans.html.

[82] Wilkinson R, Marmot M, eds. Social determinants of health: the solid facts. 2nd ed. [electronic version]. Geneva: World Health Organization, 2003. Available from URL: http://www.euro.who.int/document/ e81384.pdf.

[83] Hillemeier M, Lynch J, Harper S, Casper M. Data set directory of social determinants of health at the local level [online]. 2003. Available from URL: www.cdc.gov/CVH/library/data_set_directory/pdfs/data_set_ directory.pdf.

[84] Berkman LF, Kawachi I, eds. Social epidemiology. New York: Oxford Press, 2000.

[85] Kawachi I, Subramanian SV. Health demography. In: Poston DL Jr, Micklin M, eds. Handbook of population. New York, NY: Klewer Academic/Plenum Publishers, 2005.

[86] Kawachi I, Berkman LF, eds. Neighborhoods and health. New York: Oxford University Press, 2003.

[87] Marmot M. The influence of income on health: views of an epidemiologist. Health Affairs 2002; 21(2):31–46.

[88] Subramanian SV, Belli P, Kawachi I. The macroeconomic determinants of health. Annual Review of Public Health 2002; 23:287–302.

[89] Kawachi I, Subramanian SV, de Almeida Filho N. A glossary for health inequalities. J Epidemiol Community Health 2002; 56:647–52.

[90] Kawachi I, Kim D, Coutts A, Subramanian SV. Commentary: reconciling the three accounts of social capital. Int J Epidemiol 2004; 33(4):682–90.

[91] Kawachi I, Subramanian SV. Social capital and health. In: Anderson N, ed. Encyclopedia on health and behavior. Thousand Oaks, CA: Sage Publications, 2004; 750–4.

[92] Barthell EN, Cordell WH, Moorhead JC, et al. The frontlines of medicine project: a proposal for the standardized communication of emergency department data for public health uses including syndromic surveillance for biological and chemical terrorism. Ann Emerg Med 2002 ;39:422-9.

[93] PHIN overview [online]. Available from URL: http://www.cdc.gov /phin/overview.html.

[94] Selden CR, Humphreys BL, Yasanoff WA, Ryan ME. Public health informatics: current bibliographies in medicine, 2001-2 [electronic version]. Bethesda, MD: National Library of Medicine, 2001. Available from: http://www.nlm.nih.gov/pubs/cbm/phi2001.html.

[95] Yasnoff WA, O'Carroll PW, Koo D, Linkins RW, Kilbourne EM. Public health informatics: improving and transforming public health in the information age: topics in health information management. Journal of Public Health Management and Practice 2001;21:44-53.

[96] Health marketing [online]. 2006. Available from URL: http://www.cdc.gov/partners/ healthmarketing.htm.

[97] Oakenfull G, Blair E, Gelb BD, Dacin PA. Measuring brand meaning. Journal of Advertising Research 2000;40:43-54.

[98] About minority health [online]. 2005. Available from URL: http://www.cdc.gov/omh/AMH/AMH.htm.

[99] Institute of Medicine. Unequal treatment: confronting racial and ethnic disparities in healthcare. Washington, DC: Institute of Medicine, 2003.

[100] Centers for Disease Control and Prevention. Eliminating racial and ethnic health disparities [online]. 2005. Available from URL: http://www.cdc.gov/omh/AboutUs/disparities.htm.

[101] U.S. Environmental Protection Agency. Global earth observation system of systems (GEOSS) [online]. 2006. Available from URL: http://www.epa.gov/geoss/basic.html.

[102] AHRQ. The role of community-based participatory research: creating partnerships, improving health [electronic version]. Washington, DC: Agency for Healthcare Research and Quality, 2003. AHRQ Publication no. 03-003 7. Available from URL: http://www.ahrq.gov/research/cbprrole.htm.

[103] Harvard Medical School. Complementary and alternative medicine use by one third of U.S. adults unchanged from 1997: steady five-year prevalence points to need for more rigorous evaluation [news release] [online]. 2005. Available from URL: http://134. 174.17. 106/news/releases /1_05CAM.html.

[104] Institute of Medicine of the National Academies. Complementary and alternative medicine in the United States. Washington, DC: The National Academies Press, 2005.

[105] World Health Organization. Health impact assessment (HIA) [online]. 2006. Available from URL: http://www.who.int/hia/en/.

[106] Genomics and disease prevention FAQs [online]. 2005. Available from URL: http://www.cdc.gov/ genomics/faq.htm.

[107] U.S. Department of Health and Human Services. Mental health: a report of the Surgeon General— executive summary. Rockville, MD: U.S. Department of Health and Human Services, Substance Abuse and Mental Health Services Administration, Center for Mental Health Services, National Institutes of Health, National Institute of Mental Health, 1999.

[108] National health interview survey [online]. 2006. Available from URL: http://www.cdc.gov/nchs/ nhis.htm.

APPENDIX I
CDC ORGANIZATIONAL CHART

U.S. DEPARTMENT OF HEALTH AND HUMAN SERVICES
CENTERS FOR DISEASE CONTROL AND PREVENTION (CDC)

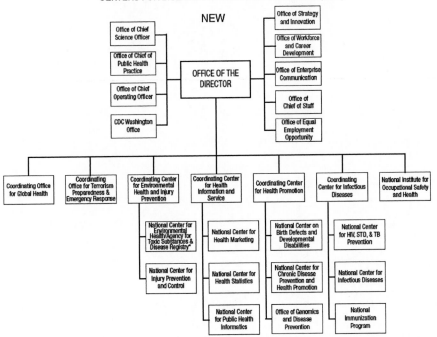

*ATSDR is an OPDIV within DHHS but is managed by a common Ofc of the Director with NCEH

Note: This graphic represents the organizational structure of CDC at the time the Research Agenda Development Workgroups were formed.

APPENDIX II
MAP OF RESEARCH THEMES TO HEALTHY PEOPLE 2010 (HP 2010) LEADING HEALTH INDICATORS AND OBJECTIVES[*]

HP 2010 Leading Health Indicators and Related Objectives	Research Guide Themes
Physical Activity	
22-2. Increase the proportion of adults who engage regularly, preferably daily, in moderate physical activity for at least 30 minutes per day.	V.D.1. Reduce the Burden of, Disparities in, and Risk Factors for Chronic Diseases Among Adults Develop, implement, and evaluate strategies to establish behaviors during adulthood that promote lifelong health and reduce the risk of the leading causes of morbidity and mortality, including tobacco use, obesity, heart disease, type 2 diabetes, chronic obstructive pulmonary disease, and cancer.
22-7. Increase the proportion of adolescents who engage in vigorous physical activity that promotes cardiorespiratory fitness 3 or more days per week for 20 or more minutes per occasion.	V.C.2. Optimal Adolescent Development Develop and evaluate strategies to improve health and fitness levels of U.S. adolescents, and establish health behaviors that promote lifelong health and reduce the leading causes of morbidity, mortality, and disability among youth and adults.

Appendix (Continued)

Overweight and Obesity	
19-2. Reduce the proportion of adults who are obese.	V.D.1. Reduce the Burden of, Disparities in, and Risk Factors for Chronic Diseases Among Adults Develop, implement, and evaluate strategies to establish behaviors during adulthood that promote lifelong health and reduce the risk of the leading causes of morbidity and mortality, including tobacco use, obesity, heart disease, type 2 diabetes, chronic obstructive pulmonary disease, cancer, injury, and violence.
19-3c. Reduce the proportion of children and adolescents who are overweight or obese.	V.C.2. Optimal Adolescent Development Develop and evaluate strategies to improve health and fitness levels of U.S. adolescents, and establish health behaviors that promote lifelong health and reduce the leading causes of morbidity, mortality, and disability among youth and adults.

*The Leading Health Indicators (left justified sub headers) will be used to measure the health of the nation over the next ten years. Each of the Leading Health Indicators has one or more objectives (left column, numbered) from Healthy People 2010 assoicated with it (1).

HP 2010 Leading Health Indicators and Related Objectives	Research Guide Themes
Tobacco	
27-1a. Reduce cigarette smoking by adults.	V.D.1. Reduce the Burden of, Disparities in, and Risk Factors for Chronic Diseases Among Adults Develop, implement, and evaluate strategies to establish behaviors during adulthood that promote lifelong health and reduce the risk of the leading causes of morbidity and mortality, including tobacco use, obesity, heart disease, type 2 diabetes, chronic obstructive pulmonary disease, cancer, injury, and violence.

27-2b. Reduce cigarette smoking by adolescents.	V.C.2. Optimal Adolescent Development Develop and evaluate strategies to improve health and fitness levels of U.S. adolescents, and establish health behaviors that promote lifelong health and reduce the leading causes of morbidity, mortality, and disability among youth and adults.
Substance Abuse	
26-10a. Increase the proportion of adolescents not using alcohol or any illicit drugs during the past 30 days.	IX.G.2. Substance Abuse Treatment and Prevention Determine the influence and impact of substance abuse on overall health and well-being for individuals and communities, and identify and evaluate best practices for prevention, assessment, and treatment of alcohol and other types of substance abuse.
26-10c. Reduce the proportion of adults using any illicit drug during the past 30 days.	IX.G.2. Substance Abuse Treatment and Prevention Determine the influence and impact of substance abuse on overall health and well-being for individuals and communities, and identify and evaluate best practices for prevention, assessment, and treatment of alcohol and other types of substance abuse.
26-1 1c. Reduce the proportion of adults engaging in binge drinking of alcoholic beverages during the past month.	IX.G.2. Substance Abuse Treatment and Prevention Determine the influence and impact of substance abuse on overall health and well-being for individuals and communities, and identify and evaluate best practices for prevention, assessment, and treatment of alcohol and other types of substance abuse.
HP 2010 Leading Health Indicators and Related Objectives	Research Guide Themes
Responsible Sexual Behavior	
13-6a. Increase the proportion of sexually active persons who use condoms.	III.E. 1. Behavioral and Prevention Research to Promote Health Develop, evaluate, and implement existing and emerging infectious disease-specific behavioral and social science interventions, public health education programs, and health communication research.

Appendix (Continued)

	V.C.2. Optimal Adolescent Development Develop and evaluate strategies to improve health and fitness levels of U.S. adolescents, and establish health behaviors that promote lifelong health and reduce the leading causes of morbidity, mortality, and disability among youth and adults.
25-11. Increase the proportion of adolescents who abstain from sexual intercourse or use condoms if currently sexually active.	III.E. 1. Behavioral and Prevention Research to Promote Health Develop, evaluate, and implement existing and emerging infectious disease-specific behavioral and social science interventions, public health education programs, and health communication research.
	V.C.2. Optimal Adolescent Development Develop and evaluate strategies to improve health and fitness levels of U.S. adolescents, and establish health behaviors that promote lifelong health and reduce the leading causes of morbidity, mortality, and disability among youth and adults.
Mental Health	
18-9b. Increase the proportion of adults with recognized depression who receive treatment.	IX.G. 1. Mental Health Determine the influence of mental health on overall health and well-being, and identify and evaluate best practices for prevention, assessment, and treatment of mental illness.
Injury and Violence	
15-15a. Reduce deaths caused by motor vehicle crashes.	VI.C.2. Risk and Protective Factors for Unintentional Injury Identify the risk and protective factors associated with the leading causes of fatal and non-fatal unintentional injury in all stages of life.

HP 2010 Leading Health Indicators and Related Objectives	Research Guide Themes
	VI.C.4. Trauma Systems Research Determine and evaluate how the components of trauma systems, including disability and rehabilitation services, improve short- and long-term health outcomes and costs for the acutely injured.
15-32. Reduce Homicides.	VI.C.1. Injury and Violence Prevention Interventions Develop and evaluate the efficacy, effectiveness, and economic efficiency of interventions to prevent and reduce the consequences of interpersonal violence, suicidal behavior, and unintentional injury.
	VI.C.3. Risk and Protective Factors for Interpersonal Violence and Suicidal Behavior Identify the risk and protective factors associated with interpersonal violence and suicidal behavior in all stages of life.
Environmental Quality	
8-1a. Reduce the proportion of persons exposed to air that does not meet the U.S. Environmental Protection Agency's health-based standards for ozone.	VI.A.2. Complex Environmental Exposures Develop, evaluate, and apply new and innovative methods for assessing the toxic action and health impact of multiple environmental exposures.
	VI.A.6. Environmental Data and Information Systems Develop, implement, and evaluate methods and tools to link available environmental hazards and health-outcome databases to support environmental public health tracking.
	IX.D.3. Health and Well-Being Across Diverse Community Settings Develop and implement strategies that enable individuals, families, schools, health systems, and employers in all communities to promote and sustain health for all of their residents.
27-10. Reduce the proportion of nonsmokers exposed to environmental tobacco smoke.	VI.A. 1 Environmental Risk Factors Establish the major environmental causes of disease and disability, and identify related risk factors.

Appendix (Continued)

HP 2010 Leading Health Indicators and Related Objectives	Research Guide Themes
Immunization	
14-24a. Increase the proportion of young children who receive all vaccines that have been recommended for universal administration for at least 5 years.	III.D. 1. Immunization Services Delivery Research Develop, evaluate, implement, and disseminate effective strategies to encourage all persons living in the United States to seek recommended vaccination, and optimize vaccine delivery.
14-29a, b. Increase the proportion of noninstitutionalized adults aged 65 years and older who are vaccinated annually against influenza and ever vaccinated against pneumococcal disease.	III.D. 1. Immunization Services Delivery Research Develop, evaluate, implement, and disseminate effective strategies to encourage all persons living in the United States to seek recommended vaccination, and optimize vaccine delivery.
Access to Health Care	
1-1. Increase the proportion of persons with health insurance.	V.A.3. Reduce Disparity in Chronic Disease Burden and Risk Factors Understand the determinants of disparities in chronic disease burden and risk factors, and implement effective interventions that address economic, structural, cultural, and individual barriers to optimal health, particularly among populations in which health disparities persist.
	IX.A.2. Social Determinants of Health Modify and evaluate social determinants of health disparities.
	IX.A.3. Health Disparities Prevention and Elimination Develop, evaluate, and promote methods and interventions that have been shown by evidence-based studies to be effective in characterizing, reducing, or eliminating modifiable health disparities.

1-4a. Increase the proportion of persons who have a specific source of ongoing care.	IX.C.2. Health-Care Delivery Systems Study effective ways to work with, collect data from, and impact health-care delivery systems.
	V.D.2. Reduce the Burden of, Disparities in, and Risk Factors for Chronic Diseases Among Older Adults Develop, implement, and evaluate strategies to establish and maintain behaviors during older adulthood that sustain health, reduce the risk of chronic disease and disability, maintain quality of life, and decrease health-care costs.
HP 2010 Leading Health Indicators and Related Objectives	Research Guide Themes
16-6a. Increase the proportion of pregnant women who begin prenatal care in the first trimester.	V.B. 1 Pregnancy Planning and Preconception Care Develop and evaluate strategies of promoting pregnancy planning and preconception care to improve birth outcomes and reduce the life-long effects of poor birth outcomes.

APPENDIX III
AGENDA DEVELOPMENT WORKGROUP
MEMBERS AND STAFF

Research Agenda Steering Subworkgroup, Workgroup on Goals and Research Agenda, Advisory Committee to the Director (ACD), CDC

Robert Galli, MD (Co-Chair, ACD Member)
Professor and Chair of Emergency Medicine University of Mississippi Medical Center Jackson, MS

Sandra Mahkorn, MD, MPH, MS (Co-Chair, ACD Member)
Chief Medical Officer for Health Information
Wisconsin Department of Health and Family Services Madison, WI

American Association for the Advancement of Science
Alan Leshner, PhD
Chief Executive Officer

American Association for the Advancement of Science Washington, D.C.
Association of American Medical Colleges Michael Whitcomb, MD
Senior Vice President
Division of Medical Education
Institute for Improving Medical Education Association of American Medical Colleges Washington, D.C.

American Medical Association
  Arthur Elster, MD
  Director of Medicine and Public Health American Medical Association
  Chicago, IL

American Public Health Association
  Harry Perlstadt, PhD, MPH
  Michigan State University
  East Lansing, MI

Association of Schools of Public Health
  Harrison Spencer, MD, MPH
  President and CEO
  Association of Schools of Public Health Washington, D.C.

Institute for Global Health
  George W Rutherford, MD
  Salvatore Pablo Lucia Professor and Director, Institute for Global Health
  University of California, San Francisco San Francisco, CA

National Business Group on Health
  Ronald A Finch, EdD
  Director, Center for Prevention and Health Services
  National Business Group on Health
  Washington, D.C.

Rand Corporation
  Nicole Lurie, MD, MSPH
  Senior Natural Scientist and Paul O'Neal Alcoa Professor of Policy Analysis
  Rand Corporation Arlington, VA

Research!America
  Mary Woolley, MA President
  Research!America Alexandria, VA

CDC Staff

Catherine A Hutsell, MPH (Deputy)
  Health Education Specialist
  National Center for Chronic Disease Prevention and Health Promotion
  CDC Coordinating Center for Health Promotion Atlanta, GA

Research Agenda Development Workgroups Participants List *

Community Preparedness and Response Workgroup

Linda Degutis, DrPH, MSN (Co-Chair)
  Associate Professor of Emergency Medicine,
  Associate Professor of Epidemiology and Public Health,
  Associate Clinical Professor
  School of Nursing
  Yale University
  New Haven, CT

Jim Lando, MD, MPH (Alternate)
  Deputy Associate Director for Science
  National Center for Chronic Disease Prevention and Health Promotion
  CDC Coordinating Center for Health Promotion Atlanta, GA

Hugh Mainzer, MS, DVM, Dipl. ACVPM (Co-Chair)
  Senior Preventive Medicine Officer, Epidemiologist Division of Emergency
  and Environmental Health Services National Center for Environmental Health
  CDC Coordinating Center for Environmental Health and Injury Prevention
  Atlanta, GA

Joe Posid, MPH (Co-Deputy)
  Public Health Advisor
  Bioterrorism Preparedness and Response Program
  National Center for Infectious Diseases
  CDC Coordinating Center for Infectious Diseases Atlanta, GA

---

* Job titles and affiliations while serving on these workgroups

Dori Reissman, MD, MPH
    Senior Medical Advisor for Emergency Preparedness and
    Disaster Mental Health
    National Center for Injury Prevention and Control CDC
    Coordinating Center for Environmental Health and Injury Prevention
    Atlanta, GA

Lisa Rotz, MD
    Acting Director
    Bioterrorism Preparedness and Response Program National Center for
    Infectious Diseases
    CDC Coordinating Center for Infectious Diseases Atlanta, GA

Neil L Sass, PhD
    State Toxicologist, Chemical Laboratory Director,
    Counterterrorism Coordinator
    Alabama Department of Public Health
    Montgomery, AL

Carol Scotton, PhD (Co-Deputy)
    Economist
    National Center for HIV, STDs, and TB Prevention CDC
    Coordinating Center for Infectious Diseases Office of the Chief Science
    Officer Health Economics Research Group Liaison
    Atlanta, GA

Elaine Vaughan, PhD
    Associate Professor, Psychology and Social Behavior School of Social
    Ecology
    University of California
    Irvine, CA

Environmental and Occupational Health and Injury Prevention Workgroup

Drue Barrett, PhD
    Acting Associate Director for Science
    National Center for Environmental Health/Agency for Toxic Substances and

Disease Registry
CDC Coordinating Center for Environmental Health and Injury Prevention
Atlanta, GA

John Corrigan, PhD (Co-Chair)
Professor and Director, Division of Rehabilitation Psychology
Department of Physical Medicine and Rehabilitation
The Ohio State University
Columbus, OH

Mindy Fullilove, MD
Professor of Clinical Psychiatry and Public Health New York State
Psychiatric Institute
New York, NY

Cindi Melanson, MPH, CHES (Deputy)
Public Health Advisor
National Center for Injury Prevention and Control CDC Coordinating Center
for Environmental Health and Injury Prevention
Office of the Chief Science Officer Public Health Education and Promotion
Network Liaison
Atlanta, GA

Daphne Moffett, PhD
Acting Deputy Associate Director for Science Division of Unintentional
Injury Prevention
National Center for Injury Prevention and Control CDC Coordinating Center
for Environmental Health and Injury Prevention
Atlanta, GA

Jeffrey Nemhauser, MD (Alternate)
Associate Chief Medical Officer
National Center for Environmental Health
CDC Coordinating Center for Environmental Health and Injury Prevention
Atlanta, GA

Lee Sanderson, PhD, MA
Senior Scientist
National Institute for Occupational Safety and Health Atlanta, GA

David Sleet, PhD
    Associate Director for Science
    Division of Unintentional Injury Prevention
    National Center for Injury Prevention and Control CDC Coordinating Center
    for Environmental Health and Injury Prevention
    Atlanta, GA

James W Stephens, PhD (Co-Chair)
    Associate Director for Science
    National Institute for Occupational Safety and Health Atlanta, GA

Elizabeth M Ward, PhD
    Director, Surveillance Research American Cancer Society
    Atlanta, GA

Craig Zwerling, MD, PhD, MPH
    Professor and Head
    Department of Occupational and Environmental Health University of Iowa
    College of Public Health
    Iowa City, IA

Global Health Workgroup

L Garry Adams, DVM, PhD, DACVP
    Associate Dean for Research and Graduate Studies Associate Director,
    Texas Agricultural Experiment Station Professor of Veterinary Pathology
    Texas A & M College of Veterinary Medicine
    College Station, TX

William C Levine, MD, MSc
    Associate Director for Science
    Global AIDS Program
    National Center for HIV, STD, and TB Prevention CDC Coordinating Center
    for Infectious Diseases Atlanta, GA

Aun Lor, MPH (Co-Deputy)
    Human Subjects Contact
    CDC Office of Workforce and Career Development Office of the Chief

Science Officer Health and Human Rights Working Group
Co-Chair and Liaison
Atlanta, GA

Jay McAuliffe, MD
Acting Team Leader, Geographic and Program Coordination Team
CDC Coordinating Office for Global Health
Atlanta, GA

David McQueen, PhD (Co-Chair)
Associate Director for Global Health Promotion National Center for Chronic
Disease Prevention and Health Promotion
CDC Coordinating Center for Health Promotion Atlanta, GA

Michael Merson, MD
Anna M.R. Lauder Professor of Public Health Department of Epidemiology
and Public Health School of Public Health
Yale University
New Haven, CT

Marguerite Pappaioanou, DVM, PhD
Professor, Infectious Disease Epidemiology Division of Epidemiology and
Community Health School of Public Health
University of Minnesota
Minneapolis, MN

Basia Tomczyk, RN, MS, DrPH (Co-Deputy) Epidemiologist
National Center for Environmental Health
CDC Coordinating Center for Environmental Health and Injury Prevention
Office of the Chief Science Officer Health and Human Rights Working
Group Co-Chair and Liaison Atlanta, GA

SV Subramanian, PhD (Co-Chair)
Assistant Professor of Society, Human Development and Health
Department of Society, Human Development, and Health Harvard School of
Public Health
Harvard University
Boston, MA

Health Information Services Workgroup

Marc Berk, PhD
>   Senior Fellow and Vice President of Health Survey Program and Policy
>   Research
>   National Opinion Research Center
>   Bethesda, MD

Lawrence Cox, PhD
>   Associate Director for Research and Methodology National Center for
>   Statistics
>   CDC Coordinating Center for Health Information and Service
>   Hyattsville, MD

Jeff Etchason, MD
>   Health Care Sector Management Lead Office of Public-Private Partnerships
>   National Center for Health Marketing CDC Coordinating Center for Health
>   Information and Service
>   Atlanta, GA

Larry P Hanrahan, PhD, MS (Co-Chair)
>   Chief Epidemiologist

Marta Gwinn, MD, MPH
>   Associate Director for Science
>   Office of Genomics and Disease Prevention CDC Coordinating Center for
>   Health Promotion Atlanta, GA

Philip Huang, MD, MPH (Co-Chair) Chronic Disease Director
>   Texas Department of Health
>   Austin, TX

Sharon Kardia, PhD
>   Associate Professor
>   Department of Epidemiology University of Michigan

Ann Arbor, MI
>   Kakoli Roy, PhD (Deputy)
>   Economist

CDC Office of Workforce and Career Development Office of the Chief
Science Officer
Economics Research Group Liaison
Atlanta, GA

Keith Scott, PhD
Professor of Psychology and Pediatrics Department of Psychology
University of Miami

Coral Gables, FL
Esther Sumartojo, PhD, MSc (Co-Chair) Associate Director for Science
National Center on Birth Defects and
Developmental Disabilities
CDC Coordinating Center for Health Promotion Atlanta, GA

Stephen B Thomas, PhD
Professor and Director
Center for Minority Health
Graduate School of Public Health University of Pittsburgh
Pittsburgh, PA

Infectious Diseases Workgroup

Terence Chorba, MD, MA, MPH, MPA, FACP
Associate Director for Science
National Center for HIV, STD, and TB Prevention CDC
Coordinating Center for Infectious Diseases Atlanta, GA
Wisconsin Department of Health and Family Services Divison of Public
Health
Madison, WI

Doresa Jennings, PhD (Deputy)
Health Communications Specialist
National Center for Health Marketing
CDC Coordinating Center for Health Information and Service
Atlanta, GA

Mary Madison, MPA
   Director, Health Services Research Blue Cross and Blue Shield Association
   Chicago, IL

Marc Overcash (Co-Chair)
   IT Specialist
   National Center for Public Health Informatics
   CDC Coordinating Center for Health Information and Service Atlanta, GA

Sabrina Walton, MSPH (Assistant Deputy) SAIC Consultant
   National Center for Public Health Informatics CDC Coordinating
   Center for Health Information and Service
   Atlanta, GA

Ruth Enid Zambrana, PhD
   Professor
   Department of Women's Studies and Director of Research Consortium on
   Race, Gender, and Ethnicity University of Maryland
   College Park, MD

Health Promotion Workgroup

Barbara Bowman, PhD
   Acting Associate Director for Science
   National Center for Chronic Disease Prevention and Health Promotion
   CDC Coordinating Center for Health Promotion Atlanta, GA

Marta Gwinn, MD, MPH
   Associate Director for Science
   Office of Genomics and Disease Prevention CDC Coordinating Center for
   Health Promotion Atlanta, GA

Philip Huang, MD, MPH (Co-Chair) Chronic Disease Director
   Texas Department of Health
   Austin, TX

Sharon Kardia, PhD
   Associate Professor

Department of Epidemiology University of Michigan

Ann Arbor, MI
   Kakoli Roy, PhD (Deputy)
   Economist
   CDC Office of Workforce and Career Development Office of the Chief
   Science Officer
   Economics Research Group Liaison
   Atlanta, GA

Keith Scott, PhD
   Professor of Psychology and Pediatrics Department of Psychology
   University of Miami

Coral Gables, FL
   Esther Sumartojo, PhD, MSc (Co-Chair) Associate Director for Science
   National Center on Birth Defects and
   Developmental Disabilities
   CDC Coordinating Center for Health Promotion Atlanta, GA

Stephen B Thomas, PhD
   Professor and Director
   Center for Minority Health
   Graduate School of Public Health University of Pittsburgh
   Pittsburgh, PA

Infectious Diseases Workgroup

Terence Chorba, MD, MA, MPH, MPA, FACP
   Associate Director for Science
   National Center for HIV, STD, and TB Prevention CDC
   Coordinating Center for Infectious Diseases Atlanta, GA

Susanne Pickering, MPH, MS, OTR/L, CHES (Deputy)
   Senior Program Management Consultant
   National Immunization Program
   CDC Coordinating Center for Infectious Diseases
   Office of the Chief Science Officer Excellence in Science

Public Health Education and Promotion Network Liaison Atlanta, GA

Richard A Schieber, MD, MPH
    Senior Medical Epidemiologist
    National Immunization Program
    CDC Coordinating Center for Infectious Diseases Office of the Chief Science
    Officer Health Research Systems Working Group Liaison
    Atlanta, GA

J Todd Weber, MD, FACP, FIDSA (Co-Chair) Director
    Office of Antimicrobial Resistance
    National Center for Infectious Diseases
    CDC Coordinating Center for Infectious Diseases Atlanta, GA

Gary Goldbaum, MD, MPH
    Senior Medical Epidemiologist, Director HIV/AIDS Epidemiology Program
    Public Health - Seattle and King County
    Seattle, WA

Bill Jenkins, PhD, MS, MPH (Co-Chair) Professor of Public Health Sciences
    Division of Science and Mathematics Public Health Sciences Institutes
    Research Center on Health Disparities Morehouse College
    Atlanta, GA

Jennie R Joe, PhD, MPH
    Professor of Family and Community Medicine and Director Native American
    Research and Training Center
    University of Arizona
    Tucson, AZ

Mark Leasure
    Executive Director
    Infectious Diseases Society of America Alexandria, VA
    Core Team
    The Core Team was comprised of the Co-Chairs and Deputies of the six
    preceding Workgroups and the Deputy of the Steering Subworkgroup plus the
    following participants:

David M Bell, MD (OSI Liaison)
  Senior Medical Officer
  CDC Office of Strategy and Innovation (OSI) Atlanta, GA

Tim Broadbent, MPA (OSI Liaison and Co-Chair) Public Health Analyst
  CDC Office of Strategy and Innovation (OSI) Atlanta, GA

Cecilia W Curry, PhD (OSI Liaison)
  Health Scientist
  CDC Office of Strategy and Innovation (OSI) Atlanta, GA

Robin M Ikeda, MD, MPH (OWCD Liaison) Associate Director for Science
  CDC Office of Workforce and Career Development (OWCD)
  Atlanta, GA

Jamila Rashid, PhD, MPH (Deputy)
  Associate Director for Policy, Planning and Evaluation Office of Minority
  Health and Health Disparities
  CDC Office of Strategy and Innovation
  Atlanta, GA

Elena Rios, MD, MSPH (Co-Chair) President and Chief Executive Officer
  National Hispanic Medical Association Washington, DC

Benedict I Truman, MD, MPH (OMHD Liaison)
  Associate Director for Science
  Office of Minority Health and Health Disparities (OMHD) CDC Office of
  Strategy and Innovation
  Atlanta, GA

Charles Williams, MPH, MA (Assistant Deputy)
  Health Communication Specialist
  Office of Minority Health and Health Disparities CDC Office of Strategy and
  Innovation
  Atlanta, GA

Dena L Williams, MPH (Assistant Deputy)
  Health Scientist, Epidemiologist
  National Center for Chronic Disease Prevention and Health Promotion

CDC Coordinating Center for Health Promotion Atlanta, GA
Office of the Chief Science Officer Staff

Darlene Spring Putnam
  Secretary
  Office of Public Health Research
  CDC Office of the Chief Science Officer Atlanta, GA

Renee Anthony
  Secretary
  Office of Public Health Research
  CDC Office of the Chief Science Officer Atlanta, GA

Cecilia W Curry, PhD*
  Health Scientist
  CDC Office of the Chief Science Officer Atlanta, GA

Cindi Melanson, MPH, CHES*
  Health Scientist
  Office of Public Health Research
  CDC Office of the Chief Science Officer
  Office of the Chief Science Officer Public Health Education and Promotion
  Network Liaison
  Atlanta, GA

Jerald O'Hara
  Extramural Liaison Specialist
  Office of Public Health Research
  CDC Office of the Chief Science Officer Atlanta, GA

Tanja Popovic, MD, PhD, F(AAM), AM(AAFS) CDC
  Associate Director for Science
  CDC Office of the Chief Science Officer Atlanta, GA

Jamila Rashid, PhD, MPH*
  Team Leader, Research Agenda and Promotion Team Office of

---

* These individuals joined OCSO after being appointed to the Research Agenda Development Workgroups and Core Team

Public Health Research
CDC Office of the Chief Science Officer
Atlanta, GA

Elizabeth L Skillen, PhD
    Health Scientist
    Office of Public Health Research
    CDC Office of the Chief Science Officer Atlanta, GA

Dixie Snider, Jr., MD, MPH
    CDC Chief Science Officer
    CDC Office of the Chief Science Officer Atlanta, GA

Robert Spengler, ScD
    Director
    Office of Public Health Research
    CDC Office of the Chief Science Officer Atlanta, GA

Robin Wagner, PhD, MS
    Associate Director for Research Planning and Evaluation Office of
    Public Health Research
    CDC Office of the Chief Science Officer
    Atlanta, GA

Guijing Wang, PhD
    Health Scientist
    Office of Public Health Research
    CDC Office of the Chief Science Officer Atlanta, GA

Trevor Woollery, PhD
    Team Leader, Strategy and Evaluation Team Office of
    Public Health Research
    CDC Office of the Chief Science Officer Atlanta, GA

APPENDIX IV: MAP OF RESEARCH THEMES TO THE OVERARCHING HEALTH PROTECTION GOALS AND CROSS-CUTTING RESEARCH

Themes	Overarching Health Protection Goals				Cross-Cutting
	People	Places	Preparedness	Global	
PREVENT AND CONTROL INFECTIOUS DISEASES					
A. Emerging and Reemerging Infectious Diseases					
1. Antimicrobial Resistance					✓
2. Environmental Microbiology of Bioterrorism-related and Other Pathogens			✓		
3. Emerging Infections and New Prevention Technologies					✓
4. Health-Care-Associated Infections and Patient Safety	✓	✓	✓		
5. Zoonotic and Vectorborne Diseases (ZVBDs)	✓		✓	✓	
B. Pandemic andS easonal In uenza					
1. Pandemic and Seasonal In uenza	✓		✓	✓	
C. Infectious DiseaseS urveillance and Response					
1. Infectious Disease Diagnostic Methods					✓
2. Infectious Disease Surveillance and Response					✓
3. Pharmacoepidemiology of Infectious and Other Disease Therapy					✓
4. Infectious Disease Elimination	✓		✓	✓	

*Research themes (which are numbered and un-bolded) are organized by their related research categories (which are lettered and bolded) and major research areas.

Themes	Overarching Health Protection Goals				Cross-Cutting
	People	Places	Preparedness	Global	
PREVENT AND CONTROL INFECTIOUS DISEASES					
D. *Vaccines and Immunizati on Programs*					
1. Immunization Services Delivery Research	✓		✓	✓	
2. Vaccine Epidemiology and Surveillance	✓		✓	✓	
3. Vaccine Safety	✓		✓	✓	
4. Vaccine Supply					✓
E. *Behavioral,S ocial, and Economic Research in Infectious Diseases*					
1. Behavioral and Prevention Research to Promote Health	✓				
2. Economic Analyses of Infectious Diseases					✓
F. *Host-Agent Interactions*					
1. Applied Genomics in Infectious Diseases					✓
2. Infectious Disease and Chronic Disease Associations	✓				
G. *Special Populations and Infectious Diseases*					
1. Health Disparities and Infectious Diseases					✓
2. Infectious Diseases Among Populations at High Risk	✓		✓	✓	
3. Perinatal and Neonatal Infectious Diseases	✓	✓	✓		

Themes	Overarching Health Protection Goals				Cross-Cutting
	People	Places	Preparedness	Global	
PROMOTE PREPAREDNESS TO PROTECT HEALTH					
A. *VulnerableC ommunitie s and Populations*					
1. Determinants of Community Vulnerability to Public Health Emergencies		✓	✓		
2. Risk Appraisal and Adaptive Behavior During a Public Health Emergency	✓		✓		
3. Predictive Strategies for Risk and Recovery in Vulnerable Populations After a Public Health Emergency		✓	✓		
4. Assessment Strategies for Populations Affected by Public Health Emergencies			✓		
5. Public Health Emergency Response Strategies			✓		
6. Public Health Emergency Management Strategies			✓		
B. *Infrastructure and Prevention*					
1. Critical Infrastructure Systems and Processes			✓		
2. Public Health and Clinical Response Systems			✓		
3. Human Migration, Mobility and Quarantine Issues Associated with Public Health Emergencies	✓		✓	✓	
4. Community Actions in Public Health Emergencies		✓	✓		
5. Local and Regional Operational Strategies for Managing Public Health Emergencies			✓		

Themes	Overarching Health Protection Goals				Cross-Cutting
	People	Places	Preparedness	Global	
PROMOTE PREPAREDNESS TO PROTECT HEALTH					
C. *Public HealthW orkforce Preparation and Front-line Prevention and Response*					
1. Community and Regional Response During Public Health Emergencies			✓		
2. Support for Front-line Personnel Involved in Health Protection Functions During Public Health Emergencies			✓		
3. Prociency of the Public Health Workforce in the Event of Disaster			✓		
D. *Detection and Diagnosis of the Hazards and MedicalC onsequencesA ssociated with Emergency Events*					
1. Health Surveillance Systems Involved in Public Health Emergencies			✓		
2. Rapid Clinical Diagnostic Capabilities During Public Health Emergencies			✓		
3. Environmental Detection and Decontamination During Public Health Emergencies			✓		
4. Rapid Assessment of Exposure and Impact Data from Public Health Emergencies			✓		
E.C *ommuni cations*					
1. Risk Communication and Information Disseminations		✓	✓		
2. Emergency Response Communications Technology			✓		

Themes	Overarching Health Protection Goals				Cross-Cutting
	People	Places	Preparedness	Global	
PROMOTE PREPAREDNESS TO PROTECT HEALTH					
F. *Community Preparedness and Response Improvement*					
1. Outcome Measurement for Preparedness Improvement		✓	✓		
PROMOTE HEALTH TO REDUCE CHRONIC DISEASES AND DISABILITY					
A. *HealthA cross the Lifespan*					
1. Implement Effective Health Promotion Strategies	✓	✓			
2. Reduce the Burden of Chronic Diseases					✓
3. Reduce Disparity in Chronic Disease Burden and Risk Factors					✓
B. *Infant and Maternal Health*					
1. Pregnancy Planning and Preconception Care	✓				
2. Conditions Associated with Hereditary Birth Defects and Blood Disorders					✓
3. Health Birth Outcomes	✓				
4. Promote Safe Motherhood and Infant Health	✓				
C. *Health and Development ofC hildren andA dolescents*					
1. Optimal Child Development	✓	✓			
2. Optimal Adolescent Development	✓	✓			

Themes	Overarching Health Protection Goals				Cross-Cutting
	People	Places	Preparedness	Global	
PROMOTE HEALTH TO REDUCE CHRONIC DISEASES AND DISABILITY					
D. *Adult and OlderA dult Health*					
1. Reduce the Burden of, Disparities in, and Risk Factors for Chronic Diseases Among Adults	✓	✓			
2. Reduce the Burden of, Disparities in, and Risk Factors for Chronic Diseases Among Older Adults	✓	✓			
E. *HealthA mong Persons with Disabilities*					
1. Health Across the Lifespan Among Persons with Disabilities	✓				
2. Early Identication of Developmental Disabilities	✓				
3. Health Among Infants, Children, and Adolescents with Disabilities	✓				
CREATE SAFER AND HEALTHIER PLACES					
A. *Environmental Health*					
1. Environmental Risk Factors	✓	✓			
2. Complex Environmental Exposures	✓	✓			
3. Environmental Biomonitoring Methods and Tools					✓
4. Environmental Health Interventions	✓	✓			
5. Lead Exposure and Health	✓	✓			
6. Environmental Data and Information Systems					✓

Themes	Overarching Health Protection Goals				Cross-Cutting
	People	Places	Preparedness	Global	
CREATE SAFER AND HEALTHIER PLACES					
B. *Occupational Safety and Health*					
1. Fatal and Nonfatal Injuries at the Workplace	✓	✓			
2. Occupational Diseases	✓	✓			
3. Occupational Musculoskeletal Disorders (MSDs)	✓	✓			
4. Safe Workplace Design	✓	✓			
5. Organization of Work	✓	✓			
6. Emerging Workplace Hazards	✓	✓			
C. *Injury and Violence*					
1. Injury and Violence Prevention Interventions	✓	✓			
2. Risk and Protective Factors for Unintentional Injury	✓	✓			
3. Risk and Protective Factors for Interpersonal Violence and Suicidal Behavior	✓	✓			
4. Trauma Systems Research					✓
5. Connection Among Multiple Forms of Violence					✓
WORK TOGETHER TO BUILD A HEALTHY WORLD					
A. *Supporting Goals for Global Health*					
1. Global Mortality Among Mothers and Young Children	✓			✓	
2. Immunization to Eliminate and Protect Against Global Diseases			✓	✓	

Themes	Overarching Health Protection Goals				Cross-Cutting
	People	Places	Preparedness	Global	
WORK TOGETHER TO BUILD A HEALTHY WORLD					
3. Global Micronutrient Malnutrition	✓	✓		✓	
4. Global Efforts Regarding Human Immunodeficiency Virus/Acquired Immunodeficiency Syndrome (HIV/AIDS), Tuberculosis (TB), and Sexually Transmitted Diseases (STDs)	✓			✓	
5. Global Water Safety		✓	✓	✓	
B. *Disease and Injury Prevention and Control in Global Settings*					
1. Global Infectious Diseases					✓
2. Global Burden of Non-communicable Diseases					✓
3. Global Burden of Injuries					✓
4. Global Occupational Health		✓		✓	
C. *Health of Vulnerable Populations in Global Settings*					
1. International Complex Humanitarian Emergencies (CHEs)	✓		✓	✓	
2. Public Health Consequences of Exploitation of Women and Children in Global Settings	✓		✓	✓	
3. Orphans and Other Vulnerable Children in Global Settings	✓		✓	✓	

Themes	Overarching Health Protection Goals				Cross-Cutting
	People	Places	Preparedness	Global	
WORK TOGETHER TO BUILD A HEALTHY WORLD					
D. *Societal Determinants of Health in Global Settings*					
1. Relationship Between Socioeconomic Status and Global Health					✓
2. Human Resources and Health Outcomes in Global Settings					✓
E. *Tools for Global Public Health*					
1. Global Measurement of Health, Disease, and Injury					✓
2. Health Marketing and Health Education in Global Settings					✓
3. Evaluation of the Effectiveness of Global Health Interventions					✓
MANAGE AND MARKET HEALTH INFORMATION					
A. *Public Health Data*					
1. Statistical and Data Science					✓
2. Data Collection					✓
3. Data Integration					✓
4. Data Analysis					✓
5. Data Dissemination					✓
B. *Public Health Informatics*					
1. Analytical Methods for Informatics					✓
2. Information and Data Visualization					✓
3. Communications and Alerting Technologies					✓
4. Decision Support					✓
5. Electronic Health Records					✓
6. Knowledge Management					✓

Themes	Overarching Health Protection Goals				Cross-Cutting
	People	Places	Preparedness	Global	
MANAGE AND MARKET HEALTH INFORMATION					
C. *Health Marketing*					
1. Informed Consumer Health Choices					✓
2. Integrated Health Marketing Programs					✓
3. Health Awareness to Health Action					✓
4. Niche Marketing					✓
5. Public Health Brand					✓
6. Message Bundling					✓
7. Emergency and Risk Communication					✓
8. Entertainment Education					✓
9. Health Literacy and Clear Communication					✓
10. Customize Health Marketing Campaigns					✓
PROMOTE CROSS-CUTTING PUBLIC HEALTH RESEARCH					
A. *Social Determinants of Health and Health Disparities*					
1. Burden of Disease					✓
2. Social Determinants of Health					✓
3. Health Disparities Prevention and Elimination					✓
B. *Physical Environment and Health*					
1. Global Climate Change					✓
2. Natural and Built Environment and Health					✓
3. Physical and Sociocultural Environment and Health					✓

Themes	Overarching Health Protection Goals				Cross-Cutting
	People	Places	Preparedness	Global	
PROMOTE CROSS-CUTTING PUBLIC HEALTH RESEARCH					
C. *Health Systems and Professionals*					
1. Public Health Systems					✓
2. Health-Care Delivery Systems					✓
3. Workforce and Career Development					✓
D. *Public Health Science, Policy, and Practice*					
1. Intervention and Translational Research					✓
2. Social, Anthropological, and Behavioral Sciences in Public Health					✓
3. Health and Well-Being Across Diverse Community Settings					✓
4. Economics and Public Health					✓
5. Community-Based Participatory Research (CBPR)					✓
6. Public Health Evaluation Research					✓
7. Use and Allocation of Pharmaceuticals and Other Medical Interventions					✓
8. Complementary and Alternative Medicine (CAM)					✓
E. *Public Health Education and Promotion*					
1. Public Health Education Research					✓
F. *Human Genomics and Public Health*					
1. Genomics Bridge Between Preventive Medicine and Public Health Research					✓
2. Human Genomics, Public Health Surveillance Systems, and Research Studies					✓

Themes	Overarching Health Protection Goals				Cross-Cutting
	People	Places	Preparedness	Global	
PROMOTE CROSS-CUTTING PUBLIC HEALTH RESEARCH					
3. Models to Incorporate Education and Community Engagement into Population-Based Genomics Research					✓
4. The Genomic Evidence Base in Public Health Practice					✓
5. Human-Pathogen Genomics					✓
G. *Mental Health and Well-Being*					
1. Mental Health					✓
2. Substance Abuse Treatment and Prevention					✓
H. *Law, Policy, and Ethics*					
1. Public Health Policy and Law					✓
2. Ethics in Public Health					✓

INDEX

E

N

S

W

Y

Z